THE POLITICS OF CLEAN AIR

EPA Standards for Coal-Burning Power Plants

Elizabeth H. Haskell

PRAEGER SPECIAL STUDIES • PRAEGER SCIENTIFIC

Library of Congress Cataloging in Publication Data

Haskell, Elizabeth H., 1942–
The politics of clean air.

1. Air—Pollution—Standards—United States.
2. Coal-fired power plants—United States. 3. United States. Environmental Protection Agency. I. Title.
TD888.P67H37 363.7'392 81-13863
ISBN 0-03-059701-3 AACR2

Published in 1982 by Praeger Publishers
CBS Educational and Professional Publishing
a Division of CBS Inc.
521 Fifth Avenue, New York, New York 10175 U.S.A.

23456789 145 987654321

Printed in the United States of America

PREFACE

This book is a case study of regulation for clean air. It documents the U.S. Environmental Protection Agency's (EPA) decision process in setting air pollution control standards for future coal-fired power plants to be built by electric utility companies. It is a story of air pollution, coal production and the utility industry, and the politics and policies of federal regulation.

The EPA experience from January, 1977 when the Agency began to review its 1971 rule, to June, 1979 when the revised standard became final, reveals the complex world of a pollution regulatory agency that must balance air quality needs with economic and energy constraints in the politically charged atmosphere of national environmental policy-making.

The political, economic, and environmental stakes were high. Coal-fired power plants are major sources of several pollutants and emitted 65 percent of sulfur oxides nationally in 1976. As the utility industry increases coal use, it presents EPA and the nation with one of its most critical environmental issues. Actors involved in this case included the president, White House staff, EPA staff and leaders, other federal departments, members of Congress, and private environmental and industry groups.

The case shows how EPA, just as any other regulatory agency, must weigh and manage these government and interest group forces. It demonstrates how regulators use mathematical models to predict effects of their standards and the limits of such technical analysis in rule-writing. It is clear from this story that personalities and the opinions of persons the president appoints to be key decision-makers in regulatory agencies are critically important. These individuals who perform the environmental/energy/economics balancing act significantly affect the outcome of environmental regulation.

EPA managers who took part in this case considered this standard to have been developed in a model process, and the case illustrates the way EPA goes about setting other national standards for air and water quality.

EPA's experience here is also an example of the challenges that other federal regulators face as demands increase for cost-benefit analyses of regulatory alternatives.

I wrote this case study in 1980 for Harvard University's Kennedy School of Government and the School of Public Health, where it is in use as teaching material in environmental public policy courses. The material has been updated for publication here.

The research involved many personal interviews with participants in the decision process, a study of the extensive EPA docket accompanying the power plant standard, and a review of related published materials.

The book is organized chronologically, in the following chapters:

1. *Introduction* provides a quick summary and issues to follow in the case.
2. *Setting the Stage* describes the legal and political background as EPA began to revise its standards, including the Congressional directives and the political alliance behind the Clean Air Act, the Carter Administration commitments about coal and air quality, the nature of power plant pollution, and key EPA decision-makers.
3. *EPA's First Position: A Uniform Standard* describes EPA's early studies to support the standard, the early draft standard, and the disputes within the Agency and from industry groups and the Department of Energy over the proposed uniform scrubbing standard.
4. *Modeling Alternative Standards* tells how EPA conducted computer-based mathematical modeling to evaluate alternative power plant standards, working jointly with the Department of Energy.
5. *The September Federal Register Proposal* tells how EPA decided to publish alternative standards under study, without endorsing either a uniform or variable standard, and reactions to this position.
6. *Resolving the Issues* tells how EPA came to select a variable scrubbing rule, resolved other regulatory issues, and secured White House agreement for its standard.
7. *Final Rule* explains the impacts of NSPS and the uproar following the EPA standards.
8. *Conclusions: Balancing on a Tightrope* evaluates the standard selected in this case and the process by which it was chosen. Lessons are drawn from this case for the analytical and political processes conducted by regulatory agencies as they go about regulating the environment, health, and safety of the nation.

CONTENTS

LIST OF TABLES

ACRONYMS

Btu	British Thermal Unit, a measure of heat
CAA	Clean Air Act
CEA	Council of Economic Advisors, three-member advisory group to the president
DOE	United States Department of Energy
EDF	Environmental Defense Fund; a public interest environmental law organization located in Washington, D.C.
EPA	United States Environmental Protection Agency
FGD	Flue gas desulfurization systems used for cleaning sulfur oxides from power plant flue gases
NAPCTAC	National Air Pollution Control Techniques Advisory Committee; air quality advisory group appointed by EPA to advise the agency on proposed rules and regulations
NRDC	Natural Resources Defense Council; a public interest environmental law organization located in Washington, D.C.
NSPS	New Source Performance Standards, required by Section 111 of the Clean Air Act
OAQPS	Office of Air Quality Planning and Standards of the U.S. Environmental Protection Agency; the organization most directly responsible for drafting New Source Performance Standards
OPE	Office of Planning and Evaluation; the office within EPA responsible for policy planning and program evaluation
ORD	Office of Research and Development; the EPA office responsible for pollution control research and development

RARG	Regulatory Analysis Review Group; a group of federal officials representing Executive Branch economic and regulatory agencies and the Executive Office of the President whose purpose was to review regulatory analyses prepared by regulatory agencies
UARG	Utilities Air Regulatory Group; an ad hoc group of the nation's electric utilities formed for the purpose of lobbying the industry's position in Clean Air Act legislation and EPA regulation

CHRONOLOGY OF THE NEW SOURCE PERFORMANCE STANDARDS (NSPS) FOR COAL-BURNING POWER PLANTS

1970	Clean Air Act (CAA) passed, requiring NSPS to be set by EPA for all major categories of sources of air pollution.
December 23, 1971	EPA promulgates NSPS for coal-burning power plants.
1973	Sierra Club challenges 1971 standards in court. Dismissed on procedural grounds.
1975	House Subcommittee on Health and the Environment drafts amendments to Section 111 of CAA requiring percent reduction of emissions, in addition to emission limit required in the 1970 Act.
1976	House passes CAA amendments, including revision to Section 111. Senate passes amendments without Section 111 changes. Conference Committee reports CAA amendments, including House language. Bill dies at adjournment.
1976	Sierra Club petitions EPA to revise NSPS.
1976	President Jimmy Carter elected.
January 1977	EPA's OAQPS announces it will review NSPS. Studies begin.
March 1977	Douglas M. Costle appointed administrator of EPA.
April 1977	President Carter announces National Energy Plan, including a requirement for "best available control technology" in all new coal-fired plants, including those that burn low-sulfur coal, meaning full scrubbing.

May 1977	EPA's OAQPS holds public hearings on NSPS revision.
July 1977	David Hawkins appointed assistant administrator of EPA for Air and Waste Management.
August 1977	Clean Air Act Amendments enacted, including revisions to Section 111, requiring a percentage reduction figure as well as emission limit for a revised NSPS that must be set within 1 year.
November 1977	Hawkins defines NSPS issues for Costle. Assumes full scrubbing standard. Expects to promulgate proposed standard by February 28, 1978.
November–December 1977	EPA's OPE, ORD opposition to full scrubbing crystalized. OPE studying partial scrubbing alternatives.
December 14–15, 1977	EPA's OAQPS presents draft NSPS requiring full scrubbing to NAPCTAC advisory meeting.
December 1977–September 1978	Phase 1 modeling of national impacts of alternative standards.
January 1978	Senator Henry M. Jackson asks for a task force to be convened to study Japanese scrubber technology.
February 1978	Georgia Power Company officials write president, Congress opposing full scrubbing.
February 1978	House Science and Technology Committee convenes hearings to study negative impacts of full scrubbing on new technology.
May–June 1978	OAQPS/OPE establish joint staffing, funding for standard-development and technical analysis.
June 15, 1978	Business Roundtable meets with president to oppose full scrubbing.
July 14, 1978	Sierra Club asks court to order EPA to propose NSPS by August 7 date set in the Clean Air Act. Court orders EPA to propose by September 12; final standard in 6 months.
August 1978	ICF firm hired by OAQPS, OPE, DOE to per-

	form remaining computer modeling analysis of alternative NSPS.
August 1978	Joint EPA/DOE task force established to analyze alternative NSPS.
August 1978	DOE pushing for partial scrubbing, ranging from 33 to 85 percent reduction of SO_2.
August 1978	Costle briefs Eizenstat, Schlesinger. Costle agrees to drop reference in NSPS proposal to "statutory presumption" favoring full scrubbing.
September 12, 1978	NSPS alternatives published in *Federal Register*. EPA indicates preference for full scrubbing, based on legalities of Clean Air Act. Phase 1 modeling results published, favoring partial scrubbing.
September–December 1978	Phase 2 modeling, using new modeling assumptions agreed upon by interagency task force.
December 1978–June 1979	Phase 3 modeling. Dry scrubbing technology incorporated.
December 8, 1978	EPA publishes Phase 2 modeling results in *Federal Register*. Emission ceiling shown to be significant.
December 12, 1978	EPA holds public hearing on modeling results and proposed standards.
January 15, 1979	RARG report favors partial scrubbing as more cost-effective.
February 1979	EPA begins microeconomic analysis of emission ceiling impacts on high-sulfur coal markets.
April 5, 1979	EPA holds meeting with National Coal Association, environmental, industry, labor groups to discuss coal reserves data and coal cleaning. NCA presents new data.
April, 1979	Costle approves partial scrubbing, based on dry scrubbing for low-sulfur coals, wet scrubbing for high-sulfur (70 to 90% reduction). Is considering 1.0 and 1.2 lb./mill. Btu emission limits, averaged monthly.

April 23, 1979	Costle, Eisenstat, NCA, others meet with Senator Robert Byrd to discuss emission limit and impacts on high-sulfur coal regions.
April 30, 1979	Costle briefs the president, administration officials on alternative NSPS, favoring 70 to 90% reduction, says emission limit not decided. No presidential directive.
May 2, 1979	Costle meets with Senator Byrd, NCA, Interior officials on coal reserves information; EPA describes emission limits it's considering.
May 3, 1979	Byrd, 25 other senators write the president opposing any lowering of the 1.2 lb. emission ceiling.
May 8, 1979	Byrd meets with the president to oppose lowering the 1.2 lb. emission ceiling.
May, 1979	Costle decides on 1.2 lb. emission ceiling.
May 25, 1979	Costle announces final NSPS.
June 11, 1979	Final NSPS promulgated in *Federal Register*.
June–July, 1979	Sierra Club, Environmental Defense Fund, Utility Air Regulatory Group, Kansas City Power and Light, California Air Resources Board, other utility groups petition EPA to reconsider the new NSPS.
February 6, 1980	EPA denies petitions for reconsideration of the NSPS. Published rejections in the *Federal Register*.
September 22, 1980	Appeals to EPA's decision argued in U.S. Court of Appeals.
April 29, 1981	U.S. Court of Appeals affirms EPA's setting of NSPS.

1

INTRODUCTION

The bright television lights clicked on as Douglas M. Costle stepped up to the podium in the crowded auditorium. Costle, administrator for the Environmental Protection Agency, was about to announce the most controversial and expensive decision made in the history of the Agency: tough air pollution control rules for new power plants burning coal; a series of regulations designed, said Costle, to make new plants five to ten times cleaner than existing ones.

That night, TV would report the new regulation, already considered the litmus test of the Carter administration's handling of environmental and energy conflicts. The newscasters would tell of the effects on coal production and rising costs of electric power and pressure by the coal lobby and then Senate Majority leader Robert Byrd.

Costle felt good about this decision. He didn't always feel as confident walking out to announce an EPA action, but this time he felt the Agency had done its homework. The rule was defensible. The numbers came from the most extensive modeling analysis EPA had undertaken to back up a new standard. Costle also had secured White House agreement for his rule. These were the kinds of support Costle knew would be needed to sustain the utility industry and environmental group challenges that were sure to come.

"These new standards," Costle said, "will help protect air quality by substantially reducing emissions from new coal-fired plants. Nationally, sulfur dioxide emissions from the new plants will be reduced to half the levels allowed by the existing standard. Particulate emissions will be 70 percent lower and nitrogen dioxide will be 20 percent lower."[1]

Costle continued, "The new standard will allow the country to move forward without disruption to fully develop coal resources while supplying energy growth needs and easing dependence on imported oil.

It will preserve our options for future growth by not allowing our clean air resources to be consumed by power plants."[2]

"Too strict, too expensive," the utilities would charge, as the new rules were projected to cost utilities an extra $35 billion by 1995. "Too lenient, threatens visibility in the West, a cave-in to industry," environmental organizations would counter.

"We guess that everybody you can think of will sue us,"[3] said David Hawkins, EPA's assistant administrator for air quality. He was right.

The New Source Performance Standards (NSPS), as the new rules were called, set limits for all new fossil fuel-fired utility boilers over 250 million Btu per hour of heat input. Plants that burn high-sulfur coals will be required to remove 90 percent of the potential SO_2 emissions before discharging gases to the atmosphere. Plants that will burn low-sulfur coal would be required to remove a minimum of 70 percent of uncontrolled SO_2 emissions. All plants would have to meet an emissions ceiling of 1.2 lb. of SO_2 per million Btu. As a result of this standard, all large new utility power plants will be required to install and properly operate some type of flue gas desulfurization system, considered by EPA to be the best available control technology, or to provide equivalent reduction through advanced technology, such as fluidized bed combustion or coal gasification.

Although less controversial than the SO_2 requirements, the NSPS also regulates particulates and nitrogen oxide (NO_X) emissions. The standards:

- Reduce allowable particulate emissions to 0.03 lb./mill. Btu from the previous standard of 0.1 lb./mill. Btu. This limit assumes utilities will install either a high efficiency electrostatic precipitator (ESP) or fabric filter (baghouse);
- Reduce nitrogen oxide emissions to 0.6 lb./mill. Btu for bituminous coal and 0.5 lb./mill. Btu for subbituminous coals from the original standard limit of 0.7 lb./mill. Btu. This limit assumes controls by modifying combustion techniques, and requires no external control device;
- A compliance with sulfur dioxide standards and NO_x will be determined by using a 30-day rolling average, with data collected by continuous monitors installed in the power plant stacks and made available to the government regulatory agencies. Compliance with particulate limits is established by manual stack tests.

Costle and his agency had taken special pains to strike a careful balance of environmental, energy, and economic needs of the nation. It was a balance that had to be achieved while walking the tightrope of national environmental and energy politics.

The *Wall Street Journal* reported that "The new, tougher, federal standards limiting pollution from future coal-fired power plants represent a careful attempt to balance competing energy, economic and regional political interests."[4] The *Washington Post* editorialized, "That was a good balance the Environmental Protection Agency struck in its rules for new power plants, though you wouldn't know it from the uproar."[5]

This NSPS struck a balance between what environmentalists claimed was technologically feasible—a uniform 95 percent reduction of SO_2 from all coals burned—and that which the utility industry advocated—a variable reduction of SO_2 ranging from 20 percent removal of potential emissions to 85 percent, with the percent removal increasing with the sulfur content of the coal.

In setting this standard for the 350 new power plants (all coal-fired) expected to be built by 1995, Congress required EPA to consider not only the technological feasibility and environmental impacts of the standard, but also its economic and energy consequences for the utility industry, electricity consumers, high- and low-sulfur coal markets, and the nation as a whole.

The utility industry, coal interests, environmentalists, Congress, the White House, Council of Economic Advisors, Council on Wage and Price Stability, and the Department of Energy were all major actors in EPA's politically charged decision atmosphere.

The political as well as economic stakes were high. EPA expects the final standards to cost the utility industry an extra $35 billion by 1995 and to raise residential electric bills by about 2 percent. But the emission reduction potential was also seen as great. The final rule is expected to reduce national SO_2 emissions in 1995 to 20.5 million tons, compared with 23.8 million tons under current rules.[6] Thus, the standard was crucial as both symbol and substance.

Perceiving that much was at stake in this NSPS, EPA Administrator Costle and other EPA managers sought to conduct the standards-setting process with extra care, committing exceptional levels of funds, staff, and top management attention. The result, the Agency feels, was a model process in environmental decision-making.

This monograph explores the process by which a regulatory agency strikes the crucial balance when setting national environmental policy: who participates and how, the role of mathematical modeling and other technical analysis, and how this is linked with policy judgments, interagency and White House politics, the importance of personalities, and personal credibility in the human business of regulation writing.

This case study also examines the substance of the regulation—where did the fulcrum lie when the balance was struck on key issues?

Table 1.1: Development of the New Source Performance Standards for Fossil-Fuel-Fired Utility Boilers

1971 NSPS	*September, 1978 Proposed Revision to NSPS*	*August, 1979 Final Revised NSPS*
SO_2		
1.2 lb./ mill. Btu heat input emission limit	1.2 lb./mill. Btu heat input emission limit	1.2 lb./mill. Btu heat input emission limit
	0.2 lb./mill. Btu emission floor	0.6 lb./ mill Btu emission floor
	uniform 90% removal of potential SO_2 emissions	70–90% removal of potential SO_2 emissions, depending on sulfur content of fuel burned
3-hour stack test to show compliance	24-hour averaging period, continuous monitors	rolling 30-day averaging; continuous monitoring
Particulate		
0.10 lb./ mill. Btu heat input	0.03 lb./mill. Btu heat input	0.03 lb./mill. Btu
	99% reduction of SO_2 emissions for solid fuels	99% reduction of SO_2 on solid fuels
	70 % reduction on liquid fuels	70% reduction on liquid fuels
Manual stack test for compliance	Manual stack test for compliance	Manual stack test for compliance
NO_X		
0.7 lb./mill. Btu on coal burned	0.50 lb./mill. Btu when firing subbituminous fuels derived from coal	0.5 lb./mill. Btu when burning subbituminous coal
	0.60 lb./mill. Btu when	0.6 lb./mill. Btu when

	firing bituminous coal	firing bituminous coal
	Continuous monitors	Continuous monitors

KEY ISSUES OF THE CASE

The following are the key issues to follow in this case study, all of which concern the SO_2 portion of the standard, which was the controversial feature.

1. *Uniform versus variable control of SO_2 emissions* (also called full vs. partial control.) The key policy issue was should EPA require power plants burning cleaner low-sulfur coals to reduce emissions by the same percentage as plants burning dirtier high-sulfur coals? Or should the percentage reduction of SO_2 vary and decrease when sulfur content decreases? If a uniform percentage reduction of SO_2 was required (such as 90 percent removal of SO_2 from all coal burned) expensive wet scrubbing systems would effectively be required on all new utility boilers that burned coal. If a lesser percentage removal were allowed when firing cleaner low-sulfur coals, cheaper dry scrubbing systems could be used, or a portion of flue gases could bypass the scrubber, saving money.

2. *Percentage reduction of SO_2.* Whether a uniform or variable standard was used, a percentage figure(s) had to be set to specify the portion of the total potential emissions that should be removed. Setting this (these) number(s) was largely a question of the best achievable efficiency of scrubbers. Environmentalists claimed scrubbers could achieve 95 percent reductions of SO_2; industry claimed 85 percent; and EPA's final rule called for 90 percent reductions for wet scrubbers on high-sulfur coals and 70 percent when using dry scrubbers on cleaner coals. The actual percent reduction slides between those two percentages, varying with sulfur content of coal burned.

3. *The ceiling for SO_2 emissions.* Ceilings of maximum allowable emissions considered in the process ranged from 0.55 lb. of SO_2 per million Btu on an annual average to the 1.2 lb. on a monthly average (about 1.0 lb. on an annual average) established in the final rule. The ceiling from the previous rule had been 1.2 lb. on an annual average. The ceiling is the controlling number for power plants burning high-sulfur coals. The percent reduction is controlling for low-sulfur coal burning. This ceiling decision was to become the symbolic issue in the

rule-writing, receiving considerably more public and interest group attention than the emissions impact of the decision warranted.

4. *The floor*, or the level of SO_2 emissions below which a utility could reduce or eliminate scrubbing. Initially, at the time of proposal, raising the floor from 0.2 to 0.5 lb. SO_2/mill. Btu was considered as a way to allow partial scrubbing, that is, to allow some part of the flue gas to bypass the scrubber, saving dollars. In the final rule, the 0.6 lb. floor, with a 70 percent minimum percentage reduction, was set to encourage and accommodate dry scrubbing.

5. *The averaging time* for calculating the SO_2 emission ceiling and the percentage reduction. For example, 90 percent reduction of SO_2 emissions averaged monthly is equivalent to 85 percent averaged daily. Averaging times considered ranged from a 3-hour period to annual averaging, with the final rule mandating a 30-day rolling average. Averaging time and continuous monitoring requirements were largely enforcement issues that had also taken into account the variability in the performance of scrubbers and the variabiity in the sulfur content of coal.

2

SETTING THE STAGE

This chapter discusses how EPA set a New Source Performance Standard for coal-burning power plants in 1971 and the prevailing legal and political environment in 1977 when EPA began to revise that standard. The nature of power plant pollution and the EPA decision-makers are described.

Congress, by way of the Clean Air Act Amendments of 1977, directed EPA to set a revised NSPS for new utility boilers that burn fossil fuels. That standard was to have two parts: a percent reduction of potential emissions and an emission limit that reflected the "best technological system of continuous emission reduction" after considering cost, energy, and non-air quality environmental impacts. How did this congressional policy directive come about? What did it mean, and what was the unusual political alliance behind this policy choice?

WHAT IS A NSPS?

This story starts in 1970, when Congress, through the Clean Air Act of 1970, directed EPA to set NSPSs limits on emissions from large new stationary sources of pollution.* These limits were to be set for categories of sources, such as utility boilers, coke ovens, and industrial

*Section 111 of the 1970 Clean Air Act required the EPA administrator to set air pollution limits for new sources to be "the best system of emission reduction which (taking into account the cost of achieving such reduction) the Administrator determines has been adequately demonstrated." [Clean Air Act Amendments of 1970, Pub. Law No. 91–604, Sec. 4 (a), Sec. III (a), 84 Stat. 1676.]

boilers. The emission limit is technology based, since it is determined by the performance of the best pollution control system available for individual categories of sources. The NSPS emission limit is the amount of pollution that would be discharged by the source, after installing the best abatement equipment. As the state-of-the-art in pollution control technology advances, the law directs EPA to revise the standard to require emissions achievable by the new technology.

While EPA is directed by the Clean Air Act to set ambient air quality standards based exclusively on health considerations, the NSPSs are to be set after EPA considers economic, energy, and non-air quality environmental considerations as well.

The aim of the NSPS is to ensure that all new sources of air pollution employ the best system of controls affordable to minimize future emissions while still providing maximum opportunity for economic growth.

The technology-based NSPS are a backup to ambient air quality standards, providing insurance that new sources of pollution will make maximum controls efforts, irrespective of ambient air quality standards or cleanup of existing pollution. The best pollution control on a plant-by-plant basis may require better control than would be shown to be necessary to meet ambient standards, as demonstrated by modeling. Furthermore, an emission or technology-based standard for each new facility is easier for state and federal agencies to include in regulatory permits and to enforce. Once written into regulations, many industries will plan for meeting the standards when building a new plant and purchase and install the equipment as a matter of course. The NSPS provides the industry with a technology-specific requirement that enables more precise capital planning.

Congress, in the 1970 Clean Air Act and the 1977 Amendments to the Act, treats new sources more strictly than existing sources of air pollution. Existing sources must meet a state-specified pollution control standard, usually expressed in terms of emission limits. The state sets these limits in their State Implementation Plans (SIPs), which are strategies to attain and maintain the national ambient air quality standards set by EPA. The SIPs set limitations for individual sources so that specific geographic areas will be cleaned up to ambient standards levels and maintained at that level. The amount of reduction by each existing source is determined by atmospheric diffusion modeling. The reduction factor is assigned to each source, so that when all sources comply with state limits, the national ambient standard will be met. Since limits for existing sources are geographic-specific, existing sources in highly polluted areas may be required to install better pollution controls than existing sources in cleaner regions.

New sources must meet a consistent and higher standard. All new sources of the same type, no matter where they are located, must meet the NSPS, at a minimum. These limits incorporate the best technology, considering cost, energy, and other environmental factors. In some clean air areas governed by the Prevention of Significant Deterioration (PSD) features of the 1977 Amendments, new sources may be required to reduce pollutants even more than the NSPS limits. These decisions are made on a case-by-case basis. In highly polluted areas not attaining ambient standards, new sources may be required to install stricter controls than NSPS to achieve the lowest achievable emissions limit.

Congress treats old and new sources of pollution differently because the economics and politics of regulating them differ. It is usually more difficult, from an engineering standpoint, and more costly to retrofit existing facilities with pollution abatement equipment than to build controls into a new plant from the beginning. Some older, less productive facilities cannot justify the cost or raise the capital.

On the other hand, most industries do not resist to the same degree installing best controls on new plants, when they have time to plan, design, and budget the equipment from the start of construction. Thus, Congress more readily approves new source controls than retrofits. Congress rarely considers, for example, whether it might be cheaper to attain clean air by controlling new power plants or existing copper smelters in the West, or whether it will produce cleaner air to control existing power plants or new ones more strictly. The tradeoffs between new and existing sources are rarely made in legislative debates. The hope is that in the future, as older, dirtier plants are closed, better air quality will be phased in and room for growth with clean air be possible.

As this case will show, the more favorable industrial attitude toward new source controls is not generally shared by the utility industry. New source controls are extremely expensive for utilities with air pollution controls averaging about 20 percent of costs of a new plant.[7] Also, much of the air pollution control equipment designed to clean combustion gas was considered unreliable by utilities.

THE 1971 NSPS

In response to the 1970 Clean Air Act directive, on December 23, 1971, EPA promulgated NSPS for large fossil-fueled steam generators operated by utilities, those with heat inputs of over 250 million Btu per hour.[8] The standard had one limit on SO_2, an emissions limit or ceiling of 1.2 lb. of SO_2/mill. Btu of heat input. No percent reduction of potential emission was specified. The standard was set based on an engineering judgment as to what constituted "the best system of emission reduction," as required by the 1970 Act. Analysis of the total impacts on the nation—costs, emissions, etc.—was not performed.

To meet the 1971 NSPS, a new power plant could burn low sulfur coal—that which contained less than 1.2 lb. of sulfur—and not install control equipment. The emissions from such burning could meet the 1.2 lb. emission limit without using any pre- or postcombustion cleaning methods. Uncontrolled emissions of SO_2 from combustion of most U.S. coals range from about 0.8 lb. to 8 lb./mill. Btu, so many cleaner coals, found largely in the Western states, could be burned without controls.

When burning high sulfur coal, the emission limit in the 1971 NSPS required flue gas desulferization systems (FGD) to remove SO_2 from the flue gases. These systems are sometimes called *scrubbers*, since they scrub the flue gases clean. The exhaust gases from the combusion process, laden with pollutants, flow up the power plant smokestack. In the scrubber these gases are sprayed with a lime or limestone solution, the sulfur dioxide in the gas reacts with the spray and goes into solution. It is later removed, dewatered, and disposed of in the form of sludge. On an average day, a scrubber will remove 190 tons of SO_2 and consumes over 400 tons of limestone and thousands of gallons of water. While these systems were improving, they were not in wide use and most utilities still considered them to be expensive and unreliable. When the 1971 standard was set, only three scrubbers were operating in the United States. The administrator determined that these scrubbers, at a removal efficiency of 70 percent, were "adequately demonstrated" technology that could be the basis for a standard. The wet scrubbing systems also produced considerable wet sludge as a waste product that had to be disposed.

To avoid buying scrubbers, utilities that could reasonably purchase, transport, and burn low-sulfur Western coal without scrubbers, such as utilities in the West and Midwest, did so. While low-sulfur coal was more expensive than high-sulfur varieties and some added transportation cost was incurred to move the coal East, the total cost was still less than buying and operating scrubbers.

As a result, the established Midwestern and Appalachian coal regions, containing mostly intermediate- and high-sulfur coals, were losing some of their market to lower-sulfur coal suppliers in Montana, Wyoming, and Colorado. Since electric utilities are by far the largest purchasers of U.S. coal, this factor had a major impact on high-sulfur coal markets. Demand for Western coal by Midwestern utilities rose from 0.1 percent of the total of 93.4 million tons purchased in 1979 to 24.7 percent of the 119.3 million tons purchased in 1977.[9] At the same time some utilities in the East switched from burning high-sulfur coal to low-sulfur oil in order to meet air pollution requirements. Thousands of coal miners were put out of work in Appalachia and the Midwestern coal fields as a result of the declining demand for high-sulfur coal. In West Virginia, by 1978, 7000 miners were unemployed and another 1500 to 2000 were working short weeks. In Ohio, 2000 miners were unemployed and another 2600 were working short weeks. In West Kentucky, Indiana, and Illinois, 1800 miners were unemployed.[10] A number of small coal companies went bankrupt. The elected representatives of these regions became increasingly concerned and began to see the 1971 NSPS as a part of their regions' economic and political problems.

Environmentalists, too, opposed the 1971 NSPS, even though ambient health-based SO_2 standards were being met in most areas. Giant power plants were being built in the West, such as those at Four Corners in the Southwest, without any pollution controls other than burning low-sulfur coals. Groups such as the Sierra Club linked emissions from these power plants with impaired visibility of spectacular vistas in the West, particularly the scenic wonders of the Grand Canyon. Environmentalists also pointed to the health risks of power plant emissions. Furthermore, environmentalists opposed, as a general principle, the construction of a major polluting source, such as a power plant, with no technological controls, when those controls were available.

In the early 1970s a few scrubbers had been installed in U.S. utilities, either retrofited on existing plants or installed in new ones. The experience with these led environmentalists to conclude that 90 percent removal of SO_2 was possible, at least when burning low sulfur coals, on which scrubbers performed better.

In 1973, the Sierra Club and the Oljato and Red Mesa chapters of the Navaho Nation, whose land was in the Four Corners area of the Southwest most heavily impacted by power plant emissions, challenged the 1971 standards in court, demanding that the standard require 90 percent removal of potential SO_2 emissions, irrespective of the type of coal burned. Such requirement would have mandated scrubbers on the Four Corners power plant. The case was dismissed on procedural grounds because the plaintiffs had not exhausted their administrative remedies.[11]

Then, in 1976, the Sierra Club petitioned EPA to revise the NSPS on the grounds that air pollution control technology had improved so that 90 percent SO_2 removal was demonstrated and thus legally required.[12] During the 1971–1976 period, environmentalists had come to view scrubbers as the necessary technology for utilities' air pollution control, while the utilities were coalescing in their opposition to scrubbers.

In January, 1977, EPA's Office of Air Quality Planning and Standards (OAQPS) announced it would review the 1971 standard. Studies were launched to determine whether a new standard was required and, if so, what should be the desired form of a new standard. Public hearings were held in Washington that May to solicit public comment. As soon as EPA announced it would review the standard, environmentalists assumed that a significant tightening of the standard would occur.

Other events and debate surrounding the 1977 Amendments to the Clean Air Act and the National Energy Plan contributed to this impression and raised the issues and interest group positions on the forthcoming NSPS revision.

CLEAN AIR ACT AMENDMENTS AND THE ENVIRONMENTALISTS/DIRTY COAL ALLIANCE

A most unusual political alliance developed during the 1975–1977 debate on amendments to the Clean Air Act that favored scrubbers. A political marriage of convenience between environmental and high-sulfur coal interests helped determine the language of Section 111 of the Clean Air Act that governs the NSPS, and pervaded EPA's setting of the NSPS.

Environmental groups sought an amendment to the Clean Air Act to mandate a NSPS that in addition to the usual emission limit would also require a percentage reduction requirement for SO_2. A percentage reduction of potential SO_2 would require control equipment and would not permit low sulfur coal to be burned without further controls. Environmental groups from their experience with the Four Corners power plants saw scrubbers as the answer to Western visibility threats—a way to prevent clean air areas from becoming as polluted as the Eastern coal-burning areas. A uniform, 90 percent reduction of SO_2 was their goal, irrespective of the sulfur content of the coal burned, because this would require FGD systems on all new power plants.

High-sulfur coal interests—coal companies and miners' unions—in the Midwest (Ohio, Illinois, Indiana, and Wisconsin) and Appalachia (West Virginia, Kentucky) wanted scrubbers for quite different reasons. A uniform national percentage reduction that mandated scrubbers would eliminate for many Eastern and some Midwestern power plants the cost advantage of purchasing higher priced low-sulfur Western coal to comply with the standard. If all power companies had to purchase scrubbers, regardless of coals burned, cheaper high-sulfur coals would be selected. A scrubber requirement would also tend to minimize regional differences in the construction and operating costs of new power plants.

Utilities were dead set against scrubbers, pointing to the early failures of these control devices, their cost, and sludge disposal problems. As alternatives, the utilities offered tall stacks, that alleviate local air pollution problems by dispersing pollutants greater distances, and by intermittent controls, in which pollutant discharges were reduced more during periods of high pollution and unfavorable weather conditions.

In 1976, after Clean Air Act oversight hearings, the House Subcommittee on Health and the Environment proposed amendments to Section 111 of the Clean Air Act calling for a NSPS that "reflects the degree of emission *reduction* achievable through the application of the best *technological* system of *continuous* emission reduction,"[13] (emphasis added) as contrasted with the previous NSPS "which reflects the degree of emission *limitation* achievable through the application of the best system of emission reduction." By changing "limitation" to "reduction" and calling for a technological system, the Subcommittee was calling for hardware that would reduce potential emissions. In the minds of House committee people, that meant universal use of scrubbers. The House committee report said a source "may no longer meet NSPS requirements merely by use of untreated oil or coal."[14] "Continuous emission reduction was added to prevent new power plants from using intermittent controls to abate pollution. These were clearly environmentalists' objectives.

The report found six problems with the existing NSPS to justify the change in statutory language. The report, embodying the following rhetoric of high sulfur protectionism, demonstrates the alliance forming between environmentalists and the high sulfur coal representatives.

- The standards give a competitive advantage to those States with cheaper low-sulfur coal and create a disadvantage for Midwestern and Eastern states where predominantly higher sulfur coals are available;
- These standards do not provide for maximum practicable emission reduction using locally available fuels, and therefore do not maximize potential for long-term growth;
- These standards do not help to expand the energy resources (that is, higher-sulfur coal) that could be burned in compliance with emission limits as intended;
- These standards aggravate compliance problems for existing coal-burning stationary sources which cannot retrofit and which must compete with larger, new sources for low-sulfur coal;
- These standards increase the risk of early plant shutdowns by existing plants (for the reasons stated above), with greater risk of unemployment;
- These standards operate as a disincentive to the improvement of technology for new sources, since untreated fuels could be burned instead of using such new, more effective technology.

The House report further assumed that the emission limit of 1.2 lb. of SO_2/mill. Btu of heat input would not change from the 1971 NSPS, which satisfied the high-sulfur coal interests. A marked reduction in this emission limit would once again freeze out the high-sulfur coal suppliers, whose coal would not be able to comply even using scrubbers.

The House committee language had the support of environmental groups, such as the Natural Resources Defence Council, and high-sulfur miners' elected representatives, and the United Mine Workers union, which represented Eastern and Midwestern miners. The National Coal Association, however, took no position, since its members included both Western low-sulfur coal companies and Eastern high-sulfur coal companies, who took opposing positions on the scrubber issue.

The full House of Representatives passed the new Section 111 in its amendments to the Clean Air Act in 1976.[15] The Senate version of the amendments included no change in the old Section 111.[16]

The Conference Committee reported a bill that included the House-passed language on Section 111,[17] but the Clean Air Act Amendments were never enacted that year. Lobbying efforts of the auto industry, which was seeking more lenient auto emission limits, the Chamber of Commerce, utilities, and Western industrial interests, particularly opposing PSD requirements, blocked passage. The auto industry figured it would receive more favorable action the following year, when Congress would be under pressure to amend the 1970 Clean Air Act before its August, 1977 deadlines forced the auto industry to halt production.

Section 111 received relatively little attention in 1976, despite the EPA estimate that costs of full scrubbing would be $14 billion over the costs of complying with PSD. PSD would likely impose scrubbing on low-sulfur coals in some parts of the West, but would not require scrubbers nationwide.

In January, 1977, after election of a new president, Jimmy Carter, and a new Congress, the Clean Air Act Amendments came up again, under increased pressure for passage by summer. The deadlines of the 1970 Act, in addition to affecting the auto industry, would, if not amended by July, require the denial of any new permits for new sources of air pollution in areas not attaining ambient air quality standards.

NATIONAL ENERGY PLAN

In 1977, the call for scrubbers on all new power plants intensified and the alliance solidified between environmentalists and dirty coal in support of the Clean Air Act.

President Carter focused the nation's attention on the energy crisis, and his National Energy Plan, drafted under the direction of James Schlesinger, the future Secretary of the Department of Energy, gave added impetus to scrubbers. The environmentally sound use of coal was the linchpin of the president's plan.

In February, the president appointed the Rall Committee, named for its chairman, David P. Rall, Director of the National Institute of Environmental Health Sciences. The committee concluded that increase in the use of coal contemplated by the Energy Plan would be safe for the environment and public health only if "universal adoption and successful operation of best available control technology on all new facilities" were required.[18]

BACT was assumed to be 90 percent reduction of SO_2 by use of flue gas desulfurization systems, or wet scrubbers. While not clearly articulated at this time, most parties were assuming that this percentage would be calculated as an annual average.

Furthermore, the new administration wanted the support of the environmental community for the president's Energy Plan and a promise of scrubbers would likely secure their endorsement. Consequently, the administration, in its fact sheet accompanying the president's energy program released in late April, said it would "require the installation of the best available control technology in all new coal-fired plants, including those that burn low-sulfur coal."[19]

Administration spokesmen testifying on the National Energy Plan, Douglas M. Costle, appointed Administrator of the Environmental Protection Agency in March, Secretary of Energy James Schlesinger, and James O'Leary, Deputy Secretary of DOE, backed "best available control technology on all new power plants."

CLEAN AIR ACT AMENDMENTS PASS IN 1977

Thus, it was no surprise that EPA recommended that Congress pass the proposed 1976 House amendments to Section 111 with its call for percent reduction and an emission limit. The House report language favoring the burning of "locally available coals" (meaning high-sulfur coal), and the percent reduction requirement would gain the support of both environmental groups and the UMW. Congressional and administration supporters of strong Clean Air Act Amendments wanted to be sure that union groups would favor passage and not ally themselves with utilities and other groups working to defeat NSPS and PSD provisions.

The House Interstate and Foreign Commerce Committee reported

H.R. 6161 in 1977, with the committee changes to Section 111 that it had proposed the previous year. The House Committee Report cited the six problems with the 1971 NSPS it had offered the previous year.[20]

Another House amendment helped cement the support of high-sulfur coal interests even further. Congressman Paul Rogers, Chairman of the House Subcommittee on Health and the Environment, responsible for the Clean Air Act Amendments legislation, offered an amendment on the floor that was adopted and became a new Section 125, "Measures to Prevent Economic Disruption of Unemployment." The section would permit any governor, the EPA administrator, or the president to find that a shift from "locally or regionally available coal" would result in "significant local or regional economic disruption or unemployment." Then the president or governor could order the source to use local coal to meet the NSPS or SIP emission requirement. This would mandate scrubbers on high-sulfur coal in these areas to meet the clean air limit.

The amendment was supported by the United Mine Workers Union, AFL-CIO, United Steel Workers of America, the United Transportation Union, as well as the National Clean Air Coalition, the Environmental Policy Center, and the Sierra Club.

In the Senate, Senator Edmund Muskie, chief Senate architect of the Clean Air Act, opposed the amendment, because it was primarily economic protectionist legislation and not clean air legislation and tended to create regional economic tensions. The amendment sqeaked by, however, by a vote of 45–44. Senator Domenici from New Mexico, opposed to uniform scrubbing, then secured this addition, "In taking any action under this subsection, the Governor, the President or the President's designee, as the case may be, shall take into account the final cost to the consumer of such an action."[21] The high cost of scrubbers would have to be considered in the decision.

The Senate, as it had the previous year, passed a bill that included no changes in Section 111.

The conference committee included Senator Domenici, who opposed the House requirement for uniform scrubbing, Senator Randolph of West Virginia who favored it, House members who favored scrubbers, and several members whose final positions were not clearly known. With many other issues considered more crucial, the conference committee did not devote major attention to Section 111. The result was language, drafted by House staffers, that was a victory for the House full scrubbing version. The bill would require that a NSPS for fossil-fuel-fired power plants include an emission limit, as before, plus a requirement to reduce a specified percentage of pollution from fuel burned. If a high percentage requirement were set—85 percent or 90 percent, for example, EPA could, in effect, make all coal-burning power plants install scrubbers. Other new sources must meet only a specific emission limit, as power plants had under the 1971 NSPS.

As the enacted language stated, the NSPS must:

> Reflect the degree of emissions limitation and the percentage reduction achievable through the application of the best technological system of continuous emission reduction which (taking into consideration the cost of achieving such emission reduction and any non-air quality health and environmental impact and energy requirements) the Administrator determines had been adequately demonstrated. . . ."[22]

It was clear from the language "technological system" and "percentage reduction" that Congress had precluded the use of untreated low-sulfur coal alone as an emission reduction system, that a percentage reduction and an emission limit would both be required.

While the language did not specify scrubbers, it certainly made them possible for the administration to require, since certain high percentage reduction figures could only be met through the use of scrubbers. On the other hand, the language of the statute does not specifically require a uniform national percentage reduction, but only "percentage reduction to be determined after the Administrator takes into consideration the cost of achieving such emission reduction and other factors."

While environmental and labor groups saw Section 111 language as a hard-won victory, the matter was made unclear by the accompanying conference report. The House and Senate conferees, with differing points of view, reached a political compromise that would allow unanimous conferee endorsement of the conference report and the legislation to permit passage by the August 5 deadline.

Utility interests, led by their lobby group called Utilities Air Regulatory Group (UARG),[23] and UARG counsel George C. Freeman, Jr., of the law firm of Hunton and Williams, had become alert to the Section 111 changes. Taking some time away from lobbying against PSD, Freeman was pushing friendly members of Congress for either no percentage reduction or at least a variable percent reduction standard, one in which the percentage reduction required would decrease as the amount of sulfur in coal declined. Freeman was working closely with Senator Domenici on the conference committee. Senator Domenici, who opposed uniform scrubbing as being too costly, taking the utility industry position, secured this language in the conference report:

> In establishing a national percentage reduction, . . . the Administrator

> may, in his discretion, set a range of pollutant reduction that reflects varying fuel characteristics."[24]

House conferees who sought a universal scrubbing requirement then added,

> "Any departure from the uniform national percentage reduction requirement, however, must be accompanied by a finding that such a departure does not undermine the basic purposes of the House provisions and other provisions of the Act, such as maximizing the use of locally available fuels."[25]

This conference language is the only reference in the act or report to a "uniform national percentage reduction," whereas Domenici's language is the only reference to a "range of pollutant reduction."

The uncertainty about congressional intent was to plague the EPA standards-setting process. While the environmentalists would point to Section 111 with its requirement for a percent reduction of SO_2 as a hard-won victory that would mandate full scrubbing, the utilities would cite the language of the conference report that provided for a range of pollutant reductions and allowed a variable standard.

Other pertinent features of the 1977 amendments included:

- EPA was to upgrade the standards to be no less stringent than existing standards;
- The NSPS was to encourage the use of local coal;
- EPA was to promulgate the revised standards by August, 1978;
- A credit may be given toward the percent reduction figure for fuel cleaning prior to combustion;
- All NSPS are to be reviewed at least every 4 years;
- The administrator could grant waivers for innovative technologies if the proposed system is not demonstrated, the system has a reasonable chance of working without risk to public health, welfare, and safety, consideration has been given to unregulated pollutants and the technology can at least meet an emission level equivalent to the applicable NSPS considering energy, cost, and environmental factors.

EPA staff, developing a revised NSPS, reading the Clean Air Act revisions, and watching the debate on the National Energy Plan, thought that the Administration was committed to full scrubbing on all new power plants—a uniform percent reduction that amounted to the "best available control technology."

POWER PLANTS AND POLLUTION: MUCH AT STAKE IN REVISING THE NSPS

As EPA began to review its 1971 NSPS, the Agency and outside parties perceived that much was at stake.

The environmental stakes were high in EPA's revision of the utility boiler NSPS. Coal-fired power plants emit sulfur oxides, nitrogen oxides, and particulate matter. The amounts vary with the quantity and quality of the coal burned, the operational procedures at the power station, and the emission control equipment applied.

Power plants emitted 18.6 million metric tons of sulfur oxides into the air in 1976, or approximately 64 percent of the total national emissions of this pollutant. These generating stations were also responsible for 29 percent of all nitrogen dioxide, and 24 percent of all particulate emissions.[26]

By 1995, if the 1971 standard for new power plants remained in effect, an estimated 23.7 million tons of SO_2 would have been emitted.[27] The anticipated growth of emissions would result from the increased demand for electricity and the shift of power generation from cleaner oil and gas to coal. In 1975, 647 million tons of U.S. coal were produced. By 1995, coal production was projected to increase in all regions of the country, and nearly triple the 1975 level by 1995, much of this going to utility boilers.[28] All new fossil-fueled power plants built in the future are expected to burn coal, because of the shortage and high cost of oil.

The environmental and health effects of pollutants from coal burning can be serious. Emissions of sulfur dioxide can irritate the upper respiratory tract and cause lung damage, particulates can cause breathing problems and respiratory illness, and nitrogen dioxide can cause bronchitis and pneumonia.

While primary ambient air quality standards for SO_2, NO_X, and particulates—those set by EPA to protect public health—are being met in most regions of the country, scientists had begun to cast doubt on the appropriateness of those ambient standards. The 1977 Act directed the Agency to review health and welfare criteria and the standards. Furthermore, the secondary substances into which these primary pollutants are chemically transformed in the atmosphere—sulfates, nitrates, and acids—are considered by some to be even more damaging than their precursors. Sulfates are thought by some scientists to contribute to major lung problems, such as asthma.

Lave and Seskin estimate that an 88 percent decrease in sulfur oxide emissions and a 58 percent decrease in particulate emissions from their 1971 levels would yield a 7.0 percent reduction in the unadjusted mortality rate. These emission reductions would also have at least as great an impact on illness or morbidity rates, they predict.[29] Despite the relationship to health, EPA would not make a health protection argument in developing its new NSPS for power plants.

SO_2 and NO_X emissions in the air contribute to the formation of acid precipitation and are thought to be killing fish in northern U.S. and Canadian lakes and may also be damaging crops and reducing farm and forestry productivity. The problems associated with acid precipitation are most pronounced in the Northeast. This is partly due to large emissions of SO_2 and NO_X from coal-burning power plants in the Midwest and East.

The chemical reaction in the air that creates secondary pollutants is thought to occur as much as hundreds of miles away from a coal-burning plant. These secondary pollutants, which are much more finely divided, are transported great distances.

Low levels of sulfur oxides, and, to a lesser extent, nitrogen oxides and particulates at levels considerably below the amount that causes public health problems, reduce visibility and impair the appreciation of scenic landscapes. Sulfates and other fine particles can scatter light, creating dull hazes. This is more of a problem in the West, where the vistas are long—up to 100 miles in places—spectacular, and air is relatively clean. When the air is cleaner, a small amount of pollution will noticeably impair visibility, whereas it will not be noticed in more polluted Eastern areas with only 25 to 35-mile views, for example.

Thus, regulation of SO_2, NO_X, and particulates is important not only because they have immediate health and property effects in the area surrounding a power plant, but also because health and environmental damage is caused by their secondary pollutants many miles away.

Just as environmental stakes were high, the costs of emission controls were also expected to be substantial. Between 1979 and 1995, if the 1971 standard had remained in effect, the utility industry would have spent about $770 billion for new power plants.[30] The addition of tighter air pollution controls would add billions, just how many was unclear. Estimates varied widely from utility spokesman to environmentalist.

As utilities pay more for fuel, construction, and transportation of fuel, electricity costs to consumers are forecast to increase dramatically over the next 15 years. The added cost of pollution abatement equipment was thus a special consumer consideration. One utility projected as much as $150 to $200 increase annually in a customer's electric bill, due to added pollution controls.[31]

While these cost estimates were tentative, they certainly were expected to be high and the financial impact of a new standard on utilities, electricity, consumers and general inflation were major concerns inside and out of government.

THE WASHINGTON ATMOSPHERE: REGULATORY REFORM

By the summer of 1977, regulatory reform had become a major feature of the Washington policy and political landscape. As the economy lagged, inflation rose, and energy shortages developed, anti-regulatory pressure came from regulated parties, in Congress, and the White House. Regulatory reform had also become a major issue within EPA, as a result of Administrator Costle's and Assistant Administrator William Drayton's concerns over costs and inflationary impacts of regulation. The Council on Wage and Price Stability (COWPS) and the Council of Economic Advisors (CEA) were becoming increasingly involved in development of regulations by EPA, Occupational Safety and Health Administration, and other regulatory agencies, pursuing what they considered inflation fighting and government efficiency improvements.

In March, 1978, President Carter created the Regulatory Analysis Review Group (RARG) to review a limited number of the regulatory agencies' analyses of proposed regulations that were especially costly. RARG is chaired by Charles Schultze, Chairman of the Council of Economic Advisors, and includes the Office of Management and Budget and representatives of principal executive branch regulatory agencies. Staff is provided by personnel of the Council on Wage and Price Stability. As a practical matter, the RARG is the main route by which the White House economists affect regulations, and their effort has been to cut the costs of environmental, safety and health regulations. EPA's revised NSPS for coal-fired power plants was to be one of their first reviews.

Environmental, labor, and consumer groups protested RARG intervention as attempts to undo major environmental, health, and safety legislation. A coalition of such groups wrote the president strongly opposing COWPS and CEA interventions, calling it "a serious threat to the interests of millions of Americans who identify with the environmental, labor, and consumer protection movements."[32] These groups felt that CEA and COWPS were not addressing the major causes of inflation, such as rising oil prices, and were concerned only with the health and safety agencies' activities, that they inappropriately substituted their judgment for that of statutes and lacked technical expertise of regulatory agencies.

EPA'S TEAM

Douglas M. Costle, whom President Carter appointed in March, 1977, to be administrator of EPA, felt that the political stakes as well as environmental and economic consequences, were very high in revising the NSPS.

Carter had chosen Costle for his political as well as his environmental judgment. Costle was known to have political antenna finely tuned from several top environmental policy jobs. He had directed the Congressional Budget Office's Division of Environment and Natural Resources and before that had been commissioner of Connecticut's Department of Environmental Protection.

From early in his EPA tenure, Costle thought the NSPS decision to be highly sensitive and important for the environment and one of the key issues for him as administrator and for the Carter administration.

Costle favored technical analysis to show costs and benefits of alternative decisions and point the direction to the right policy conclusion. He also knew that solid research had political and legal value to support an agency decision against its challengers in the executive branch, Congress, interest groups, and the courts. These participants had all become regular actors in EPA's decision process by 1977.

Washington environmental groups and the utility industry were looking to this NSPS decision as a crucial first test of the new Carter administration.

Carter came into the presidency committed to environmental protection, but the ensuing energy crisis and the need to develop coal were seen as competing presidential commitments. The NSPS would test the administration's balancing of these objectives.

The environmental community had supported Carter during the presidential campaign, and several environmental group leaders had been appointed to key administration positions in the White House, executive departments, and EPA.

David Hawkins, the head of the air quality project of the Natural Resources Defense Council, a Washington-based public interest law firm, was appointed by Carter to be EPA's assistant administrator for air and waste management in mid-1977. Hawkins was considered to be a determined and effective environmentalist, trained in the tactics of environmental advocacy law. The NSPS decision would have to meet with his approval as well as Costle's.

Another key performer in the NSPS decision process was Walter Barber, EPA's deputy assistant administrator for air quality planning and standards. Barber had chief staff responsibility for drafting the standard. Barber, with a master's degree in both engineering and public administration, had been a budget examiner for the Office of Management and Budget and directed the energy policy staff in EPA's Office of Planning and Evaluation, In January, 1977, he was selected to direct development of air quality regulations in Durham, N.C. With a reputation as a fair and tough-minded analyst, Barber was expected to emphasize the air quality program's analytical capacity and to link the Durham staff more closely with EPA's analytical unit in OPE. OPE is EPA's organization charged with long- and short-range policy planning and evaluation of existing and proposed programs and regulations. The air program staff had been located in Durham prior to EPA's creation in 1970, and some Washington staff had felt Durham employees were out of touch with Washington politics and interest groups.

Barber's counterpart in planning and evaluation, Roy Gamse, deputy assistant administrator, a policy analyst with an economist's perspective, was also to figure significantly in the analytical process that supported the NSPS decision.

Gamse worked for William Drayton, assistant administrator for planning and management, a close associate of Costle's since the days he had worked for him in Connecticut as a consultant for McKinsey and Company. Drayton, a Harvard College and Yale Law School graduate, stressed the need for cost-effective regulations and an agency-wide standards development process.

3

EPA'S FIRST POSITION: A UNIFORM STANDARD

This chapter describes EPA's first working draft—a uniform scrubbing standard—as the Agency began to review its 1971 power plant standard. Early studies to support a revision are highlighted, along with the disputes within the Agency, and outside pressure from industry groups and the Department of Energy over the proposed uniform scrubbing standard.

EPA's Office of Air Quality Planning and Standards (OAQPS), the unit that would first draft the rule, felt that the Clean Air Act presumed that the same percentage reduction of SO_2 would be required of low- as well as high-sulfur coal, unless costs, energy impacts or non-air quality considerations were unreasonable. Furthermore, the administration had committed itself in the National Energy Plan to the best technological controls on all new power plants.

EARLY STUDIES

Based on EPA's analyses of a proposed new standard by the fall of 1977—studies which focused primarily on a model plant basis—the economic, energy, and technological impacts of a uniform or full scrubbing standard were not expected to be excessive. In fact, the costs of reducing 90 percent of sulfur dioxide from low-sulfur coal was cheaper for an individual plant than removing the same percentage from high-sulfur coal, since scrubbers perform better on low-sulfur varieties. The question, then, as EPA staff saw it, was to determine what percentage SO_2 could reasonably be reduced from high-sulfur coal, and that number would become the uniform percentage reduction required on all coals.

Alternative standards being considered included reduction a uniform 90 percent removal of SO_2—considered the best available, a uniform 80 percent removal of SO_2 which would also require scrubbers but at a lower efficiency, and a 0.5 lb. emission ceiling.

At this juncture, Walter Barber and OAQPS thought the standards-setting would require a moderate EPA investment. Economic and engineering studies would be needed to determine the efficiency of scrubbers and associated costs and energy impacts.

To examine the impacts of these three alternative standards, EPA had contracted in early 1977 for the following studies to cover the issues that OAQPS thought were important to the NSPS revision:

- Mitre/Mitrek Division — Coordination or contractor studies and environmental impact statement;
- Aerotherm — Factors that affect uncontrolled SO_2 emissions from steam generators;
- Aerospace — FGD solid waste characteristics and disposal;
- ICF — Impact on coal production;
- Radian — FGD energy consumption, FGD water consumption and effluents; FGD availability;
- GCA — Classifications of modifications and reconstructions;
- Battelle — Coal cleaning and coal availability.
- Bechtel — FGD design and operating parameters, reheat systems;
- PEDCo — FGD vendor capabilities, cost analysis, FGD systems;
- Teknekron — Analysis of issues and nationwide impacts, analysis of economic impacts.

In the past EPA had assessed alternative standards primarily on a model plant basis. The costs, emissions, and other impacts of alternative standards were assessed and the most effective chosen. The 1971 NSPS had been selected on this basis, for example. The revised NSPS was the first new source standard that was subject to national impacts analysis by computer modeling.

The Teknekron analysis was based on a computer model of the utility industry, developed in large part with funding from EPA's Office of Research and Development. The model predicted electric utility capacity mix nationwide for alternative standards and SO_2 emissions from electric utilities nationally and by regions, for the three alternative standards. (See Appendix 1 for some results of this modeling.)

ICF, Inc., a Washington consulting firm, was employed by EPA to estimate production of various coals as a function of alternative new standards and alternative growth scenarios. These data were fed into the Teknekron model, run by the California consulting firm.

The various studies estimated impacts for:

- Air quality, SO_2 emissions, and ambient impacts;
- Energy consumption. FGD controls require energy to reheat the flue gases, operate the draft fans, motors, and pumps. Regenerable systems have higher energy penalities to recover the sulfur;
- Water quality (impacted by thermal wastes from cooling waters, ash handling systems which use water to make ash slurry, and general service water);
- Water quantity (FGD systems use substantial quantities of water for slurry);
- Land use (land for power generation facilities and for disposal of wastes);
- Primary economic considerations (impacts on utility to purchase and operate control equipment);
- Secondary economic impacts (primarily impacts on the coal industry and coal miners);
- Fuel mix by utilities (coal, oil, gas used).

OAQPS EARLY DRAFT NSPS

OAQPS had developed a preliminary draft NSPS which was used for discussions within EPA. It contained these features:

SO_2

- A 90 percent uniform removal of potential SO_2 on all power plants, based on use of flue gas desulfurization technology. This percent removal figure is the controlling number for utilities burning low-sulfur coal, since a lower percent removal would meet the 1.2 lb. emission ceiling when sulfur is low. Credit would be permitted for removing sulfur by coal washing prior to combustion of coal in a fluidized bed containing sorbent material, called a *solvent refined coal process*. As much as 5 percent of the 90 percent reduction of SO_2 could be achieved by coal cleaning, the remaining 85 percent coming from FGD systems;
- An emission limit, or ceiling, of 1.2 lb./mill. Btu, the same as the 1971 NSPS. No source may exceed this amount, which becomes the controlling figure for high sulfur coals. For some very high sulfur coals, the percent removal would have to exceed 90 percent in order to meet the 1.2 lb. emission ceiling;
- A 0.2 lb./mill. Btu emission floor. Any source which could meet this

emission limit would be excused from the percentage reduction requirement, but since no American coals could be burned without scrubbing and meet this emission limit, the floor, in effect, required full scrubbing or scrubbing on all coals.

Particulates

- A 0.03 lb./million Btu, based on the installation of high efficiency electrostatic precipitators (ESPs) or fabric filters (baghouses), along with FGD systems, which also remove some particulate. The existing NSPS set a 0.10 lb./million Btu emission limit for particulate;

NO_X

- Emission limits vary with the fuel (coal, gas, or oil) and are based on the removal efficiencies achieved with operational controls. No control equipment should be required to achieve these levels;

Monitors

- Continuous monitors would be required to measure SO_2, opacity, and NO_X emissions. The monitoring data could be used in court to demonstrate compliance or noncompliance with the NSPS limits. There are no continuous monitors for particulates, but opacity monitors help show particulate levels. Compliance with particulate limits would be demonstrated by performance tests—manual stack tests;

Averaging Time

- The 24-hour averaging time would be used to calculate the emissions as contrasted with an annual averaging period for the 1971 NSPS. The 24-hour averaging period was selected to satisfy enforcement requirements. The Clean Air Act had set penalties on a daily basis for violations of standards and a 24-hour averaging period would allow daily penalties to be assessed. An emission standard must specify not only the maximum amount of discharge, but also the period of time for which that amount would be calculated to determine compliance. Emissions from a source will vary over time, due to variable pollution control equipment performance and differing fuel composition. For example, sulfur content of coal varies not only from mine to mine, but from seam to seam within the same mine. A utility purchases coal of a specified quality, for example, such as having a defined ash content and sulfur content, but those parameters will vary when the load is delivered. A short averaging time, such as the 24-hour period, permits less discharge over the specified level. The power plant operator does not have time to adjust pollution control equipment as quickly as those coal and performance parameters change. Additional scrubber capacity may be required to compensate for the lack of time to adjust controls. As the averaging period lengthens, to 30 days or a year, for example, the source may be able to compensate for equipment and coal variations. The utility may have an opportunity to achieve "credit" from times when emissions are less than allowed toward periods of poor performance in the same averaging period. Thus, as the averaging period shortens, the percent reduction figure can go down, or the emission ceiling can increase and produce the same emissions. Table 3-1 shows how the averaging periods change the percent reduction and emission figures;

No malfunction or Bypass Allowed

- No malfunction of control equipment would be permitted. Any malfunction that produced excess emissions would be considered a violation. The 1971 NSPS stated that a utility plant operator could allow flue gases to bypass the pollution control equipment, increasing emissions during a malfunction, as long as a good-faith effort was being made to restore controls and public health was not endangered. This had been difficult to enforce. Malfunction requirements are related to averaging periods. If the averaging period is longer, the operator has several days, or even months, to produce reduced emissions and compensate for the high emissions during malfunction. If the averaging period is shorter, and no malfunctions specifically permitted, utilities may be required to add extra scrubber modules to control system design or to increase reserve margins in order to have capacity available during scrubber malfunction. The spare module adds cost. Also, depending on the level of maximum emission limit, a no-malfunction policy and a very short averaging period could push utilities to purchase medium- or lower-sulfur coals to assure compliance during variation in control equipment performance.

By the summer of 1977, as results of various studies began to come in, OAQPS established a working group that included representatives of other EPA programs with potential interest in the power plant standard. The Standards Revision Working Group was made up of representatives from Office of Research and Development (ORD), Office Of Planning and Evaluation (OPE), Division of Stationary Source Enforcement (DSSE), Office of General Counsel (OGC), solid waste and water quality staff, with the air quality program represented by Don R. Goodwin, Director of the Emission Standards and Engineering Division (ESED) within OAQPS in Durham, North Carolina.

A working group is an EPA requirement to elicit Agency-wide

Table 3.1: Conversion of Emission Ceilings as a Function of Averaging Periods

(Pounds per Million Btu of SO_2)		
24-Hour Averaging Period	*30-Day Average*	*Annual Average*
2.0	1.4	1.2
1.8	1.2	1.0
1.6	1.0	0.9
1.4	0.92	0.8
1.3	0.8	0.7
1.0	0.63	0.55

Source: Conversation between the author and John Crenshaw, OAQPS, EPA, August 13, 1980.

views on a regulation under development. This group met occasionally to review drafts and study findings. For example, 31 persons from Washington and OAQPS in Durham met in Durham to secure agreement on the draft standard.

By November 1977 meetings began in Washington for representatives of the various Assistant Administrators (AAs). Assistant Administrator for Air and Waste Management, David Hawkins, wanted to smooth and accelerate the AA's review of the proposed standard, just as soon as it was drafted, so as to be able to propose a standard in the *Federal Register* by February 28, 1978.

Assistant Administrator Hawkins, who had worked hard to get Congress to enact Western visibility protections in the Clean Air Act amendments that summer of 1977, including the Section 111 requirement for percentage reduction of emissions, laid out the NSPS issues as he saw them to Costle in a November 21, 1977, memorandum.[33] Nowhere in this memo does Hawkins present for consideration the choice of a variable percent control (partial scrubbing) standard in which the percent reduction figure would vary. Hawkins considered uniform reduction of SO_2 a legal and environmental imperative. The NSPS issues to be resolved, he said, were:

- "A percentage reduction" and emission limit for SO_X;
- System availability—what part of the time will scrubbers be available for pollution abatement, rather than down for maintenance or malfunction. Hawkins said that utilities could buy spare scrubber modules and power from other companies during times of malfunction, so that a power plant could operate without a bypass;
- Averaging time for calculating compliance;
- Emission limit for NO_X—"The best boiler design in terms of NO_X emissions is only available from one company," adversely affecting the other three suppliers;
- Percent reduction for particulates and NO_x, which Congressman Paul Rogers indicated would be required, would be difficult to set for particulates and impossible for NO_x, since NO_x control is accomplished through boiler design and there is no "potential" emission rate;
- Applicability of the Standard. The draft NSPS applies to utility steam generators of more than 250 million Btu per hour of any fuel, not industrial boilers, since they are covered by another standard, unless they generate electricity for sale on the grid.[34]

OFFICE OF PLANNING AND EVALUATION CONCERNS

As soon as the draft NSPS began to take specific, written form in October, opposition began to surface within EPA. The Office of Planning and Evaluation (OPE), which is EPA's policy analysis unit, expressed concern with the draft NSPS. OPE's main objection was with the full scrubbing feature of the regulation. The 0.2 lb./mill. Btu floor in the draft standard would require scrubbers to be installed on all power plants, even those burning low-sulfur Western coal, at substantial cost. A higher floor,such as 0.5 lb./mill. Btu would allow very low-sulfur coals, those less than the 0.5 lb. floor, to be burned without scrubbing. Consequently, it was called a partial scrubbing standard. A power plant emitting less than the floor 0.5 lb. would be exempt from the percent reduction requirement. OPE, through its Energy Policy staff, began to conduct its own studies to explore the implications of raising the floor to allow partial scrubbing.

Thomas Schrader of OPE's Energy Policy Staff, commenting on the actions of an October 1977 EPA working group meeting, said:

> Concerning the lower emission limits for SO_2, I do not believe that the Working Group agreed that the same SO_2 control requirement should be applied to both high and low sulfur coals. There are two basic ways of considering cost affordability and cost effectiveness. While the affordability approach considers primarily the capability of the industry to purchase and operate control equipment, the cost effectiveness approach compares costs with emission loadings. The point I was attempting to make at the meeting is that different controls on different coals *may* be justified when one considers the cost of control in light of the reduction in emission loadings. As indicated in the November 1 memorandum, OPE will analyze the cost effectiveness and other aspects of alternative lower limits (italics mine).[35]

Other OPE concerns included the 24-hour averaging period, which it considered too restrictive and likely to raise the cost of utility compliance, because backup control systems would need to be purchased to take over if equipment malfunctioned or sulfur content of coal varied. OPE contended that

> to achieve a 90 percent removal on a continuous 24-hour basis, scrubbers would have to be designed with an average 24-hour removal of over 100 percent. For a 500MW scrubbing system with modules designed for an average removal of 95 percent and with two backup

> modules, over 40 percent of the days would fall below 90 percent removal. Additional spare modules could be added, but costs would become excessive. The plant would have to shift load, shut down, or bypass (the scrubber) during these periods.[36]

OPE also sought a detailed clarification of the malfunction policy, the conditions under which the Agency would permit bypass of a malfunctioning control system, such as permitting bypass only during power emergencies. This policy would have significant impact on how utilities plan and operate plants.

ENFORCEMENT VIEWS

The enforcement division favored the 24-hour averaging period, which would show daily violations and allow EPA to assess daily noncompliance penalties, as provided for in the Clean Air Act. Continuous monitors, with their data used to establish legal compliance or noncompliance, was another enforcement objective. In the past, EPA had required continuous monitors be installed on major stationary sources, but the data could not be used in court to demonstrate compliance. The enforcement division hoped to incorporate the continuous monitoring requirement in all NSPSs and other rules to streamline enforcement proceedings.

OFFICE OF RESEARCH AND DEVELOPMENT CONCERNS

The Office of Research and Development (ORD) recommended a particulate standard of 0.05 lb./mill. Btu rather than the proposed tightening to 0.03 lb. that OAQPS was considering. ORD contended that Western power plants could comply with the 0.05 standard by using combined SO_2 and particulate scrubbers, but that to reach the 0.03 limit would require the installation of an additional electrostatic precipitator or fabric filter. Western power plants could comply with the 0.05 standard at considerably less cost than the 0.03 standard, producing a 2.5 percent savings in the cost of a new power plant.[37] ORD, as well as OPE, recognized that the 0.05 standard would allow greater fine particulate emissions in the West, but that the visibility and health implications were unknown. As Deputy Assistant Administrator for Planning and Evaluation, Roy Gamse, with oversight of OPE, put it,

I am concerned about the possible cost burden for uncertain benefits. A 0.05 standard still reduces emissions (over present standard) by a factor of two, and separate PSD and visibility regulations can be used to require additional control where necessary to protect identified air quality values.[38]

Gamse was referring to the PSD requirement that a major new source of emissions in a clean air area must install Best Available Control Technology, which is defined on a case-by-case basis. The NSPS limits establish minimum control, but BACT limitations may require additional reduction of pollutants of individual plants after the case-by-case examination.

ORD was also concerned that the 90 percent reduction of SO_2, particularly from high-sulfur coal, was not achievable, preferring instead 85 percent removal efficiency as being more economical and technologically achievable.

EPA GOES PUBLIC

As most Americans were preparing for Christmas, on December 14 and 15 in 1977, Don Goodwin of the Emissions Standards and Engineering Division was presenting the NSPS draft to the first public forum, EPA's National Air Pollution Control Techniques Advisory Committee (NAPCTAC). This 16-member committee, set up by EPA to review the draft standards and solicit public opinions, consisted of nine spokespersons from industry, six from state and local air pollution control agencies, and one person from a citizens' air pollution control organization. In addition, many interest groups and individuals from business and industry, the environmental community consumer groups, and state and local organizations were invited to attend and participate in the hearings.

Goodwin's presentation strategy was to "go public" with the toughest defensible standard, theorizing that the final rule would be negotiated with affected interests to a more moderate conclusion. The draft NSPS that had been circulating for 2 months was considered by Goodwin to be the toughest standard.

As Goodwin put it in a memo to Walter Barber:

Decisions on policy issues and on technical issues where we had little or no data were made with a philosophy of selecting the most stringent option. We followed this philosophy in order to stimulate comments and with the understanding that we may have to revise the decisions before proposal.[39]

Another consideration in presenting the toughest possible standard was stated by Goodwin's assistant, George Walsh:

> We should be pushing technology. The only mechanism we have is that of public discussion of very stringent standards. When we go out on a limb, it's like putting lead weights around the ankles of a basketball player during practice sessions! It almost seems that the final standards are unimportant; it is the discussions and the arguments about achieving a stringent standard that cause technology to change.[40]

The standard OAQPS presented to NAPCTAC was the draft described earlier, a standard based on uniform 90 percent reduction of SO_2 on all power plants or full scrubbing.

UTILITY INDUSTRY POSITION

At NAPCTAC, the Utility Air Regulatory Group opposed EPA's proposed standard on every ground.

UARG is an ad hoc group formed for the purpose of lobbying the industry's position in Clean Air Act proceedings in the legislative and executive branches. It consists of the Edison Electric Institute, the National Rural Electric Cooperative Association, and 63 individual utility systems that collectively own a majority of U.S. generating capacity. At this time of the NAPCTAC meeting, TVA was also a member. It was later to drop out of the organization when David Freeman, a supporter of pollution controls, became Chairman of TVA. Freeman disagreed with George Freeman, UARG's counsel, who continually opposed EPA standards-settings.

These were UARG's concerns:

- "Emission levels are more important than the percentage reduction of sulfur content. Once the emission levels are determined it should not be important what percentage removal the utility employs to obtain that level.
- "UARG believes the percentage removal should be set on the basis of a declining percentage removal with declining sulfur content of the coal and based on cost effectiveness. As the sulfur content decreases the cost of cleaning increases, as a result cleaner fuels should not have to be cleaned as much as dirty fuels.
- "The EPA proposal discourages a number of "front end" systems which can

remove from 75 to 85 percent of the sulfur in the fuel. (Particularly fluidized bed combustion and solvent refined coal.) The UARG proposal attempts to encourage these and other innovative technologies. UARG believes these technologies have potential for a much greater reliability than FGD systems.

- "UARG further believes scrubbers are not substantially proven to operate without some defined provision for bypass."[41]

Other problems UARG had with the EPA draft included:

- EPA did not address the waiver provision of the Clean Air Act which allows for waivers for the use of undemonstrated, new technology;
- EPA's definition of "commence construction" would make the NSPS apply to power plant construction in a number of cases already committed to compliance with existing standards. Construction and operations of these plants would be delayed and electric supply problems result;
- Ninety percent SO_2 removal has not been demonstrated to be reliable;
- Opposed EPA's disallowance of cleaning credit for SO_2 removed in bottom ash or fly ash and pulverizers that are not designed specifically for sulfur removal;
- Baghouses and ESPs to control particulates to 0.03 on boilers have not been demonstrated and could not meet EPA's limit;
- NO_x standards cannot be achieved. Corrosion of tubes is an insurmountable problem.

As an alternative, UARG proposed the following standard:

- An emissions ceiling of 1.5 lb./mill. Btu rather than EPA's 1.2 lb.;
- A sliding scale or variable standard, where the percentage of SO_2 reduction would decrease as lower-sulfer coals were burned. A standard allowing less than 90 percent removal for low-sulfur coals would allow some flue gas to bypass the scrubber (hence the term "partial scrubbing") in order to avoid the extra cost of reheating the flue gas to obtain good plume rise;
- A scale to slide from a 0.4 lb. emission floor to the 1.5 lb. ceiling;
- A malfunction allowance and a 30-day averaging time;
- A credit for sulfur removed in pulverizers or sulfur retention in the fly ash or bottom ash, as well as credit for sulfur removal in systems designed specifically for that purpose.[42]

Figure 1 was displayed by EPA at the NAPCTAC meeting to show the difference between the UARG and EPA proposals.

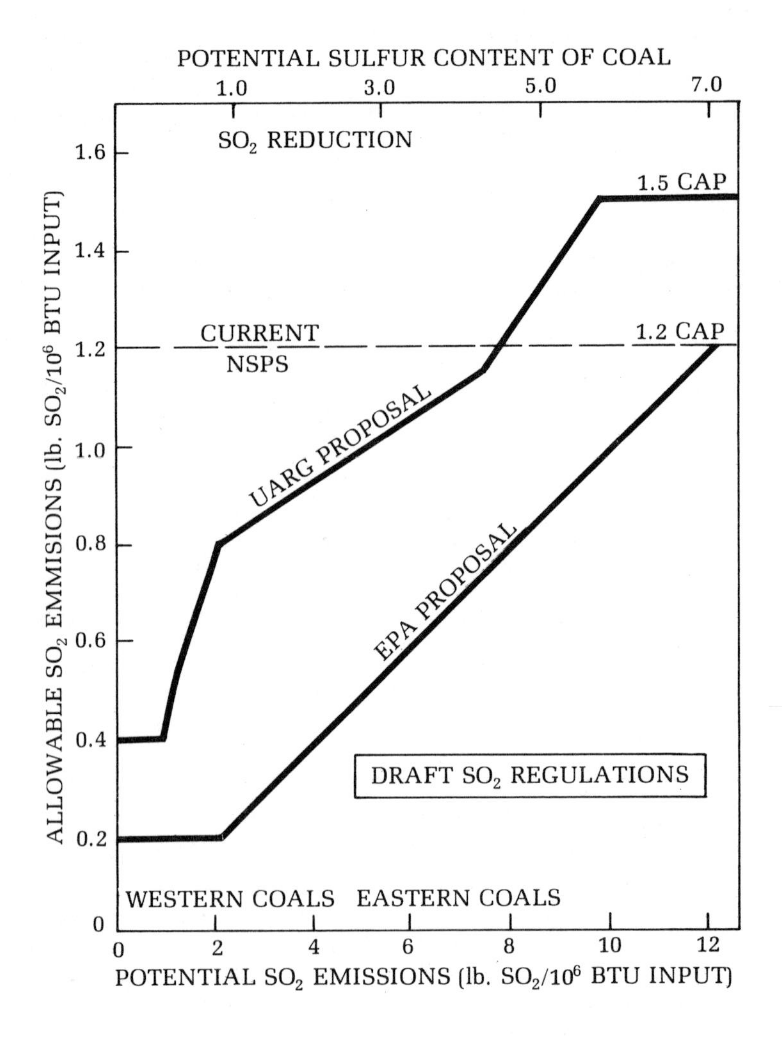

Figure 1. Comparison of EPA's Draft NSPS with Utility Industry Proposal.

Source: NAPCTAC Minutes, p. 35.

An environmentalist wrote his position for the NAPCTAC record:

Maximum Control Level. We are particularly concerned with the

> suggestions that plants burning very low sulfur coal could escape with only 85 percent FGD (the removal required before cleaning credits of 5 percent added to total the 90 percent removal called for in the draft). Scrubbers work better and cheaper on low sulfur coal. Scrubbing low sulfur coal uses less energy . . . and produces less solid waste. . . . Even UARG agrees . . . that scrubbers work better on low sulfur coal. Thus, if very low sulfur fuel should be subject to separate treatment, at all, a *higher* removal percentage seems indicated.
>
> *Particulate Standard.* Many power plants routinely achieve particulate removals of much more than 99 percent.[43]

In referring to the UARG position paper, the author concludes, "In view of all these indicia of bias, plus the obvious self-interest of the authors of the Status Report, EPA ought to treat that document for what it is: a self-serving effort to avoid or delay the mandate of Congress that 90 percent SO_2 removal must be achieved."[44]

Much of the NAPCTAC discussion focused on how well FGD systems could perform. The TVA spokesman, Gerald Hollinden, concluded that the 85 percent SO_2 reduction averaged daily was possible, putting him within 5 percent of EPA's draft 90 percent reduction standard. What troubled the utilities the most was the full scrubbing required on all coals.

POST-NAPCTAC: DEBATE BEGINS IN EARNEST

Once EPA had stated publicly its working draft of the NSPS, the utility industry, stimulated by UARG, began to lobby actively in Congress, the White House, and the Department of Energy against the proposed full scrubbing standard.

On February 3, 1978, the president of the Georgia Power Company, R. W. Sherer, in Atlanta, wrote his fellow Georgian in the White House, to protest the proposed NSPS. Copies were provided to the Georgia congressional delegation, members of the House and Senate, EPA's Costle, DOE's Schlesinger, the governor of Georgia, and members of the Georgia General Assembly. Sherer said to the president,

> A factor which deserves our leaders' urgent attention is the negative effect which the proposed Regulations for New Source Performance Standards as required by the 1977 Clean Air Act will have upon our country's utilization of its only abundant fuel resource, coal, to provide the energy needs of our nation as outlined in the National Energy Act. . . . The proposed regulations fail to take into account the economic impact and other considerations as required by the amend-

ments. . . . Therefore, Mr. President, I urge you to direct the Environmental Protection Agency to promulgate revised New Source Performance Standards in a manner which would not force unnecessarily expensive and unreliable technology to be installed in our coal-fired electric generating plants. . . . It seems clear from the draft regulations that costs and energy requirements have not been given adequate serious consideration.

For instance, the proposed sulfur dioxide emission standards may require a 90 percent removal of SO_2 independent of the sulfur content of the coal burned. The effect of this regulation would be to mandate the use of flue gas desulfurization systems (scrubbers) for new coal-burning facilities. It is clear from this choice that the EPA has gone far beyond the wishes of Congress and has based its standards on the most extreme capabilities of the technology presently showing the highest removal of sulfur dioxide, independent of the cost involved. Industry studies have shown that continuous reliable scrubber equipment operation has not been demonstrated.

We agree fully with the necessity of clean air, but we cannot support any action which places on our customers an additional cost burden without a commensurate benefit. While estimates of that additional cost vary, I believe we can reasonably expect a $150–200 increase per customer in the average annual electric bill as the result of the proposed regulations."[45]

That same day W. Robert Worley of the Georgia Power Company wrote Georgia's Senator Herman E. Talmadge,

Unless EPA adopts reasonable regulations, we greatly fear that our country never will achieve any appreciable position of energy independence because of the counterproductive restrictions on the utilization of coal. . . . If EPA forces the installation of flue gas desulfurization equipment (scrubbers) in coal fired electric generating plants, which our industry believes to be an unrealistic air quality standard, then our consumers and your constituents will experience a substantial increase in the price of electricity. . . . We would be most grateful if you would write to the President as soon as possible and send copies of your letter to Dr. Schlesinger and Mr. Costle, urging them to closely coordinate the goals of EPA in achieving reasonable clean air standards in a manner that would permit our nation to have ample electrical energy needed to continue a growing and viable economy which provides the jobs for its citizens.[46]

The Business Roundtable, a group of influential business persons and industrialists, also lobbied against EPA's draft standard. A memorandum was prepared to Robert S. Strauss, then U. S. Ambassador handling European trade talks, who sent it on to Costle. The Roundtable urged adoption of a variable percentage removal

> Such as those proposed by the electric utility industry and the Department of Energy. . . . A variable percentage standard . . . would diminish the inflationary impact of compliance, reduce the added costs to consumers, lessen consumption of imported oil, result in less scrubber sludge wastes and encourage the development and use of alternate "front-end" technologies.[47]

James Evans, Chairman of the Union Pacific Corporation and Chairman of the Environmental Task Force of the Roundtable, wrote and met with Stuart C. Eisenstat, Assistant to the President for Domestic Affairs and Policy, to transmit much the same message,

> EPA is now considering a single fixed percentage standard which virtually dictates the use of scrubber technology . . . would have the inflationary effect of maximizing the costs of compliance. It would also discourage investment in alternate emerging technologies which are economically and environmentally preferable to scrubbers. A variable percentage standard would diminish the inflationary impact of compliance, reduce the added costs to consumers, lessen consumption of imported oil, result in less scrubber sludge wastes and encourage the development and use of alternative "front-end" technologies.[48]

On June 15 the Business Roundtable met with the president to convey this message.

Throughout 1978 and up to the time the final standard was published, George Freeman of the law firm of Hunton and Williams served as UARG's counsel and lobbyist. He met with and wrote to Stuart Eisenstat, urging adoption of a variable percentage removal standard, refuting environmentalists' claim that health effects were at stake, refuting the position of the Natural Resources Defense Council that a variable percentage standard was prohibited by the Clean Air Act. Freeman also wrote the Justice Department, charging that the proposed NSPS NO_X standard violated antitrust policy, since only one of the five utility boiler manufacturers could clearly meet the revised standards for NO_X, a charge EPA rejected. Freeman lobbied actively on Capitol Hill and generated congressional support for the variable scrubbing standard. He also worked with DOE, CEA, and COWPS. Freeman was in regular written and phone communication with EPA and his written positions fill hundreds of pages in the docket of the NSPS proceedings.

THE DEPARTMENT OF ENERGY OPPOSES THE EPA DRAFT STANDARD

Before NAPCTAC, EPA had thought that the Department of Energy (DOE) would not object strongly to its draft of the revised NSPS. After all, the president and Secretary Schlesinger had committed themselves to "Best Available Control Technology," which effectively meant scrubbers, in the National Energy Plan. In May, 1977, Schlesinger had testified on Capitol Hill in favor of 90 percent uniform scrubbing. But by late 1977 DOE was backing away from its former commitment to BACT.

As EPA's Walt Barber put it, "We misjudged DOE's concern, which was an irrational concern. If electricity costs go up with a tighter NSPS, DOE should have favored it, since this would reduce the demand for electric power. DOE and Deputy Secretary O'Leary in particular were just responding to the utilities' interests."[49]

The utility industry and UARG worked with O'Leary who opposed FGD systems. In November, 1977, O'Leary had stated, "We don't have a demonstrated technology for the scrubbing of Eastern coals,"[50] at a time when EPA was saying that it had demonstrated the effectiveness of scrubbing on high sulfur Eastern coals with lime/limestone FGD systems. O'Leary was to remain an opponent of the EPA scrubbing standard throughout the rule-writing process.

By April 1978 DOE was stating the utility industry's position publicly. Dr. James Liverman, Acting Assistant Secretary for Environment at DOE, in testimony before the House Science and Technology Committee, said:

- "85 percent or higher SO_2 removal or a 24-hour basis has not yet been adequately demonstrated for either high or low sulfur coal. We are concerned that such a requirement may be pushing technology beyond its capabilities;
- the economic impact of requiring full scrubbing on low sulfur eastern and western coals may be unacceptable. Both EPA and DOE are working to identify the magnitude of this impact so that a reasonable decision can be reached which reconciles energy and environmental concerns;
- the proposed particulate removal requirement (0.03 lb./mill. Btu) could force the use of baghouses on plants using western coal. Baghouses have not yet been demonstrated for large-scale power plant applications, and such a requirement at this time could be pushing the technology beyond its capabilities . . . ;
- inclusion of certain emerging energy technologies under EPA's proposed NSPS may inhibit needed private investments for development, demonstra-

tion, and commercialization, which could prevent the gathering of necessary information on the economic and environmental performance of full-scale plants."[51]

But just as there were divergent opinions about a draft uniform scrubbing standard developing within EPA, so there were differing views within DOE.

Jim Speyer, Director of DOE's Office of Coal and Utility Policy, previously had worked in EPA where he had directed energy policy analysis for OPE. Speyer helped convince DOE that it should take greater part in EPA's process to ensure adequate consideration of energy and economic impacts in standards-setting.

Secretary Schlesinger and Assistant Secretary Alvin Alm, who had been assistant administrator at EPA for Planning and Evaluation before coming to the Department of Energy, both favored an objective technical analysis of alternative NSPS. Alm stressed cooperative dealings with EPA. Alm met with DOE's assistant secretaries to argue that the alternative standards be analyzed objectively and that interest group rhetoric not dominate the review of proposals. DOE officials were in regular communication with EPA about the draft standard, working with Drayton, Gamse, and OPE. These informal dealings were to become formal by August 1978 when a Joint EPA/DOE task force was established to analyze alternative NSPS. (This task force is described below.)

CAPITOL HILL OBJECTS

Concerns about the proposed standard were beginning to be heard from Capitol Hill, much of it stimulated by the Department of Energy and the utility industry. The House Committee on Science and Technology held hearings on February 9 and again in April on the impact of the proposed NSPS on emerging energy and pollution control technology.

A key Hill opponent of full scrubbing was Senator Henry M. Jackson, Chairman of the Senate Committee on Energy and Natural Resources. In January, Jackson convened a task force to visit Japan and evaluate scrubber use there, where technology was more advanced than anywhere else in the world. Representatives of EPA and the Electric Power Research Institute, the research organization for the utility industry, composed the task group. This group reported that Japanese scrubbers perform better than what was considered maximum efficiency in the United States. The team found

> FGD technology is working well in Japan in both utility and industrial applications. Each of the eleven scrubber installations visited was designed for and routinely attained SO_2 removal efficiencies in excess of 90 percent (93 percent average) while achieving operabilities exceeding 96 percent (98 percent average).[52]

The environmentalists were subsequently to use this finding to document a higher percent removal figure than EPA had proposed, but the finding did not change Jackson's position. By summer, Jackson's view was stated in a letter to the president, also signed by Senator Clifford Hansen from the low-sulfur coal state of Wyoming. They expressed:

> Grave concern and distress regarding (EPA's proposed full scrubbing) regulations. This proposal does not contain any provision reflective of varying fuel characteristics thus necessitating full scrubbing on all but the lowest sulfur fuels . . . the impact of these draft regulations could well be devastating. . . . It is imperative that any New Source Performance Standards for fossil fuel fired stationary sources adopted by the EPA depart from a uniform national standard in order that the hard fought gains of national energy legislation not be sacrificed for what are, in the main, non-environmental objectives. We urge that you move with all due speed in bringing EPA's action on this matter into accord with America's energy needs and this Congress' demonstrated intent (parenthetical phrase mine).[53]

In a letter to Administrator Costle, Jackson and Hansen were joined by six other Senators in opposing a uniform percent, full-scrubbing standard. The letter made the arguments that the utility industry and the Department of Energy were repeatedly making on Capitol Hill and in the executive branch.[54]

Other low-sulfur coal state politicians objected, since uniform scrubbing would reduce demand for low sulfur reserves, as compared to variable scrubbing. The governor of Utah, Scott Matheson, wrote Costle supporting DOE's positions for a variable standard and higher emission floor, citing the congressional intent to recognize varying polluting characteristics of coal, the language of the 1977 Clean Air Act Amendments conference report.[55]

UNIFORM VERSUS VARIABLE SCRUBBING DEBATE ACCELERATES WITHIN EPA

Meanwhile, back at Waterside Mall, where EPA is headquartered, debate about the draft NSPS was intensifying among EPA divisions, primarily on the issue of uniform vs. variable scrubbing. OPE and Deputy Assistant Secretary Gamse led the group favoring variable or partial scrubbing by raising the emissions floor from 0.2 lb./mill. Btu (full scrubbing) to 0.5 lb. emission floor (partial scrubbing).

Gamse summed up his partial scrubbing position in a February 28 memo to Walter Barber:

- Partial scrubbing would save money and oil over full scrubbing and result in only "moderate changes in emission loadings."
- Teknekron impacts analysis did not use "present value" analysis but only costs to 1995, which understated the full economic impact of alternative standards. "Teknekron estimates show a 1986–1995 cumulative savings of only $3.5 billion by raising the floor to 0.5 lb. from 0.2 lb. of SO_2/mill. Btu. But the actual savings realized by the utility may be considerably higher when calculated over the life of all the plants constructed during that period. Present value cost figures show a $9.3 billion savings from raising the floor to 0.5 lb. and $21.2 billion savings from raising the floor to 0.8 lb."[56]

He reiterated previous concerns

- That the 24-hour averaging period was too short; that the 0.03 lb./mill. Btu particulate standard be raised to 0.05; and that the malfunction policy be uniform and not case by case.

Gamse submitted Table 3-2 to show the impacts of partial scrubbing as compared to full scrubbing.

Full scrubbing advocates, led by Assistant Administrator David Hawkins and OAQPS, pushed for the draft NSPS, with 90 percent SO_2 removal, averaged daily, on all coals by setting the floor at an all-inclusive 0.2 lb./mill. Btu. Hawkins was also asking his staff for analysis of ceilings lower than the 1.2 lb. SO_2/million Btu contained in the draft standard.

Hawkins summed up the full scrubbing arguments in a memo to Deputy Administrator Barbara Blum. Cost of full scrubbing was not considerably greater than partial but emissions would be substantially less.*

---

*The numbers are slightly different than Gamse's since they come from slightly later technical analysis.

Table 3.2: Implications of Permitting Less than Maximum Control on Lower-Sulfur Coals

			Change Due to Different Floors for SO_2 Control Alternatives		
	Actual 1975	*Present NSPS*[a]	*0.2 lb.*	*0.5 lb.*	*0.8 lb.*
Present value cost[b] (Billions of 1977 $)	NA	36.3	+23.7	+14.4	+2.5
Price of electricity in 1990 (mills./kWhr in 1977 $)	31.5	34.0	+0.5	+0.3	+0.1
Emission loadings in 1990[c] (mill. tons/year)	18.6	21.3	−2.5	−2.0	−0.9
Oil consumption in 1990 (mill. barrels/day)	3.1	3.2	+0.4	+0.1	+0.1
Coal Production in 1990 (mill. tons/year)					
National	650	1770	−55	−15	+10
East	400	440	+30	+25	−25
Midwest	150	300	+75	+50	+10
West	100	1030	−160	−90	+25

[a]1.2 lb. of SO_2/mill. Btu and SIPs where more stringent.

[b]Includes all capital expenses and capacity shifts expected to occur as a result of alternative NSPS through 1990. Operation and maintenance expenses are calculated for the life of all plants constructed between now and 1990. Present value estimates are calculated with a 5 percent real discount rate.

[c]Assumes 90 percent average annual removal with a 1.2 lb. of SO_2/mill. Btu as a maximum emission limit. The floor is the emission level at which the 90 percent rule no longer applies.

Source: Memo from Roy Gamse, Deputy Assistant Administrator for Planning and Evaluation to Walter Barber, Deputy Assistant Administrator for Air Quality Planning and Standards, February 28, 1978.

Background

- A coalition of environmentalist and Eastern and Midwestern coal interests in the House were successful in their efforts to have the Clean Air Act amended in 1977 to require more stringent controls on new coal-fired power plants. This requirement had the dual purpose of minimizing emissions from increased coal use and maximizing the use of locally available, higher-sulfur-content coals. It also reflects the philosophy that the best control systems should be applied to new sources in order to minimize future problems and to maximize the potential for economic growth;

- These amendments were adopted in the midst of the energy policy debate, and the Congress was aware of their likely economic and energy implications. The requirements set forth in the legislation were also consistent with the administration's promise that best available control technology would be applied to all coal-fired plants as a condition of the National Energy Plan;
- Sulfur oxide control can be accomplished in a number of ways. The best demonstrated method is flue gas scrubbing with lime or limestone. This method results in significant reductions in emissions of sulfur oxides, but has attendant economic costs and energy penalties;
- Scrubbers perform more effectively on low-sulfur coals than on higher-sulfur coals. Therefore, the same level of control on plants burning low-sulfur coal would not result in any greater burden in terms of costs or energy requirements than those incurred by plants burning higher-sulfur coal;
- All new power plants must meet PSD requirements, including case-by-case review, to determine best available control technology. A less stringent NSPS for low-sulfur coals could result in protracted debate as to whether plants meeting NSPS requirements also satisfied the PSD requirements.

Impacts

Emissions. Under full control, national SO_2 emissions will be reduced by 2.5 million tons in 1990. Under partial control, the amount of reduction achieved will be reduced by 400 thousand tons;

Coal. Under full control coal production in the East and Midwest will increase by some 100 million tons. This gain will be offset by a less rapid increase in Western production. Despite this, Western production will still increase by some 800 percent over 1975 levels. Under partial control, about 40 million tons of low-sulfur Western coal will be shipped eastward, in part displacing coal that would have been produced in the East and Midwest. This would reduce the number of new East and Midwestern coal mining jobs by 6–12,000.

Costs. Under full control, a less than 1 percent increase ($700 million) in annualized cost has to be borne as compared to partial controls. Adoption of full control will only result in 19 cent increase in the national average monthly residential bill. Under full control, a reduction in cumulative capital expenditures (1976–1990) of $6 billion will result, since fewer new coal-fired units will be built.

Oil. Oil consumption due to increased use of existing oil-fired units and the addition of new turbine units will increase by 200,000 barrels per day over partial control.

Anticipated Reaction

- If a decision is made to require less than full control, aggressive adverse reaction is anticipated from virtually all environmental interest groups and from members of Congress who are particularly concerned about environmental protection and preservation of local coal markets. They will argue that partial control does not reflect best technology and its adoption will perpetuate the economic advantage currently enjoyed by those regions of the country having abundant supplies of low-sulfur coals. In addition, since all power plants are subject to the prevention of significant deterioration requirements, the affected interest groups will most likely resort to the case-by-case review provision and/or litigation as a means of bringing about full control on all plants. The resulting delays could prove costly and adversely affect implementation of the National Energy Plan.

- On the other hand, a decision to require full control will be opposed by the electric power industry, Western coal interests, and those concerned over increased costs of environmental control. In arguing against full control, these interest groups will stress the associated higher costs and energy penalties. In addition, they will point out that the higher emissions resulting from partial control are not significant at the national level and that local or regional air quality problems can be more appropriately addressed by means of the prevention of significant deterioration requirements or State regulations. They will argue further that greater use of Western coal will not result in a loss of jobs for Eastern or Midwestern miners, since coal production as a whole will increase.

- Finally, they will stress that the primary concern of the House was to preclude the use of untreated low-sulfur coal alone as a means of compliance.[57]

4

MODELING ALTERNATIVE STANDARDS

This chapter discusses the computer-based mathematical modeling sponsored by EPA and the Department of Energy to predict the impacts of alternative power plant NSPSs. The type of model and the contracting firm are described, along with the EPA and DOE staffing of the technical analysis, and the results of the first of three phases of modeling.

By the latter part of January, as utility industry, DOE, and congressional opposition to EPA's draft NSPS intensified, and EPA offices were divided on the proposed standard, Costle, Hawkins, and Barber realized that the target date of February 28 for proposing the revised NSPS could not be met. Costle had been reviewing NSPS issues since November and realized that the standards development process was becoming much more protracted than EPA staff had first expected.

EPA leaders agreed that the Agency would need to conduct further technical analyses and resolve intra-Agency disputes to reach agreement on the preferred, most defensible NSPS. OPE's concerns, in particular, would need to be addressed. Furthermore, what was, by now, an obviously controversial standard would need strong technical documentation.

SELECTING THE MODEL AND MODELERS

Following NAPCTAC, OAQPS continued to fund Teknekron's computer-based analysis of national impacts of alternative standards. OPE funded its own computer modeling, conducted by the Washington firm ICF, to show how partial scrubbing compared to full scrubbing. ICF used the kind of "net present value" analysis of total costs that OPE preferred, whereas Teknekron did not. Teknekron was showing very little difference in cost between full and partial scrubbing, but a major difference in emissions—conclusions that favored full scrubbing. OPE felt that the California firm underestimated the cost differences between partial and full scrubbing.

Hoff Stauffer, the ICF analyst, had previously worked at EPA's planning and evaluation division and now directed ICF's work on its coal and electric utilities model. ICF had developed this model for the Department of Energy's predecessor, the Federal Energy Administration. Since early 1977, EPA had been using the ICF model to provide coal impacts data to the Teknekron comprehensive national impacts model. Conclusions from the ICF model were shipped to California and fed into the Teknekron computer there. ICF continued to provide these data, while also providing more comprehensive impacts numbers to OPE.

The ICF model could predict impacts of alternative NSPS on coal markets nationally and regionally, and on utilities' expansion plans and costs on an industry-wide basis. The model forecast coal production, consumption, and prices, given such input parameters as electricity growth rates, nuclear capacity, and oil and gas prices. It was designed to show the effects of alternative NSPS on these forecast variables. The model used 30 supply regions, 35 demand regions, and six consumption sectors and 40 coal types. The model forecast production in geographic regions by coal type. It predicts (a) new capacity additions by power plant type; (b) scrubber capacity by level of partial scrubbing and by sulfur removal efficiency; and (c) utility consumption of oil, gas, and coal. It reflects the effect on new power plant capacity, scrubber capacity, and utility fuel consumption of additional costs imposed on the new coal-fired power plants by alternative NSPS.

Effects predicted included:

- SO_2 emissions nationally and by four regions;
- Utility oil consumption;
- Costs to utilities—annualized costs, present value increases, utility capital requirements, new capacity;
- Cost to electricity consumers—national average rate increases and increases by four regions;
- Coal production—nationally and by four regions, how much Western coal consumed in the East, coal prices and consumption; and
- Incremental cost per ton of SO_2 removed.

The model is a "least cost" model, which generates equilibrium solutions through a linear program formulation that balances supply and demand for coal at minimum cost. Since the ICF model is a cost-minimizing economic model, it assumes that a utility's decision is based solely on least cost. In meeting demand, it determines the most economic mix of plant capacity and electric generation for the utility system, based on a consideration of construction and operating costs for new plants and variable costs for existing plants. It also determines the optimum operating level for new and existing plants. The least cost model implies that all utilities base decisions on lowest costs and that neutral risk is associated with alternative choices.

The ICF least cost model was more responsive to cost than the Teknekron simulation model and showed greater cost savings for partial as compared to full scrubbing than the Teknekron model showed.[58] Thus, the ICF model made partial scrubbing look "most favorable." "Most favorable," it must be remembered, depends on the point of view and values of the person reading the numbers.

OAQPS throughout this process tended to be more concerned about emission reductions, whereas OPE tended to think cost reductions and reduced oil consumption were most important.

With ICF's national impacts numbers, OPE was even more convinced that partial was better than full scrubbing. Partial scrubbing would save the utilities and their customers money, reduce oil consumption, and result in only moderate emission increases. (See the numbers in Gamse's February memorandum on page 43, which were provided by ICF.)

At this point OAQPS favored full scrubbing and felt that the ICF results might be just the effect of a model anomoly. The same numbers were interpreted differently, by various parties, depending on what they considered to be most important.

JOINT OAQPS/OPE STAFFING

By summer, Walter Barber, the air program's principal manager charged with development of the regulation, and Roy Gamse, of Planning and Evaluation, took two steps to pursue their divisions' interests jointly. First, Barber and Gamse each assigned staff from his division to direct the technical analysis of alternative NSPS. Barber assigned John Haines and John Crenshaw from OAQPS, and Gamse assigned David Shaver, an OPE policy analyst, to devote full time to the regulation's development. These joint OAQPS/OPE personnel became the central staff for developing the rule. Others within EPA, such as

ORD and Enforcement, who wished to contribute to the final standard, did so through Haines, Crenshaw, and Shaver.

The formal working group, the intra-agency review body composed of staff from various offices throughout EPA, did not meet again after this.

CONTRACTOR SELECTED

In a second move, Barber and Gamse agreed to hire jointly the Washington firm, ICF, to perform all remaining computer analyses of NSPS alternatives. Gamse and, by August, Barber, had come to think that the Washington firm could deliver the required information faster than its California competitor. EPA was by then feeling the pressure of deadlines to complete the standard quickly. By using one firm EPA reasoned, the ICF coal allocation material did not have to be put into another computer, thus saving time and eliminating a possible mismatch between the two systems. Furthermore, ICF could perform "net present value" analysis and Teknekron could not, according to EPA managers. Since DOE had sponsored development of the ICF model, Barber also figured that there would be less debate with DOE over this model's use, which proved to be the case. In most other respects, the two models were similar. While the ICF model tended to show partial scrubbing to be more favorable, and Teknekron's to show full scrubbing as being slightly more favorable, this policy significance was not the basis of the switch to ICF. More intangible factors were important. For example, EPA technical staff believed the ICF model was "more rational," "better documented," and made the case for the preferred alternative in a more complete and convincing way. The Teknekron model, which was in earlier stages of development, "seemed to have dangling ends" and was not "consistent" or as persuasive in its conclusions.[59]

JOINT EPA/DOE TASK FORCE

By August, as ICF was being selected to perform all modeling analyses, EPA was also formalizing the DOE/OPE contacts by establishing a Joint EPA/DOE Task Force to analyze alternative NSPS. This key development reflected the importance that Costle, Hawkins, Barber, and DOE leaders gave to resolving their differences on technical issues. Costle and Schlesinger directed their staffs to secure agreement on the modeling assumptions (scrubber costs, oil prices, scrubber efficiency, growth in electricity demand, etc.) and modeling techniques. DOE funded half of the ICF contract costs, in an unusual joint funding with EPA.

Dan Badger of DOE's Office of Coal and Electric Utility Policy was assigned full time to work with EPA's Haines, Crenshaw, and Shaver. Badger, a Kennedy School of Government graduate, considers himself an environmentalist, just as do the EPA members of the task force, but all are quick to add that pollution control standards must be cost-effective. All shared an analytical approach to problems, had policy analysis backgrounds, and maintained free and easy communication. The task force members' compatibility helped build interdepartmental cooperation at the staff level and to resolve the differences over the NSPS analysis—a cooperation that the policy leaders of the two affected governmental organizations had mandated.

The aim of policy leaders and subsequently of the joint technical staff was to conduct analysis and arrive at technical conclusions that all parties would accept. Modeling assumptions were developed with the goal that they should be the most defensible and objective, and not clearly favoring either partial or full scrubbing. For example, EPA's staff describe Dan Badger as arguing for a position on technical grounds and not the clear utility position espoused by some DOE officials. Up to this point in August, when the joint task force was established, there had been no such agreement by DOE, EPA, and outside parties on background and impact numbers. Numbers which EPA offered lacked credibility, with environmentalists as well as industrial parties thinking the model was "stacked" to arrive at a predetermined conclusion.

Consequently, it was a conscious strategy by Costle, Hawkins, and Barber to secure agreement on factual matters and key input assumptions so that the final debate—in public and with the White House, should it reach there—would focus on policy issues. White House review of EPA's preferred standards was virtually a certainty by August, because of the magnitude of the NSPS costs, the intensity of interest group feeling about the standard, and the major political consequences of the decision.

EPA's goal was to structure the final debate among Costle, cabinet secretaries, presidential advisors, and, if necessary, present it to the president in policy terms, such as whether billions of dollars in added scrubber costs were more important than additional tons of SO_2 emissions, or whether reducing sulfur dioxide emissions in the West was more important than protecting coalminers' jobs in Eastern high-sulfur coal states. Schlesinger and Alm also wished the debate to be structured along such lines.

Typically, in previous standards setting processes, final White

House review focused on whose cost estimates one believed, or which agency had the better modeling techniques for predicted impacts of alternative standards. These technical discussions diffused the policy issues. Such technical matters, Costle believed, should be resolved by the regulatory agency and not White House staff, who had neither the time nor the technical expertise to make such judgments. On the other hand he felt that the proper White House role was to review the fundamental national policy decisions inherent in the Agency's setting of a standard. It would be EPA's responsibility to see that the debate was framed in those policy terms, and, EPA hoped, the weight of the debate would favor EPA's conclusions.

In the September–December 1978 period following publication of the NSPS proposal, representatives of the Council on Wage and Price Stability and the Council of Economic Advisors joined the Joint EPA/DOE Task Force. Meetings were usually open to industrial, coal, and environmental representatives as well as to other private individuals. In addition, utility industry people would call the EPA/DOE analytical staff about once a week to learn about the latest modeling results. Environmentalists also became deeply involved in the modeling process, analyzing assumptions, offering new ones, suggesting scenarios to be put to the model. It was becoming clear to all parties by the summer and fall of 1978 that the modeling conclusions and the importance Barber, Costle, and others attached to them were to be major factors in this decision.

Early in the standards development process, Barber and George Freeman of the utility industry had agreed to be free and open on the technical facts. As DOE came into the process, a similar pact was made with DOE. EPA also sought to include environmental groups on a regular basis. As new drafts of a standard and new modeling results became available, EPA circulated these to all parties for comment. In the last days of the standards development, this review process was to break down as court-ordered deadlines approached.

PHASE 1 MODELING RESULTS

During the 18-month period that ICF worked for EPA and DOE,* over 33 alternative NSPS were tested on the model. Some alternatives were posed by EPA, some by DOE, two by environmental groups. The

---

*ICF worked for EPA's OPE from December 1977 through August 1, 1978, and jointly for OPE, OAQPS and DOE from August 1978 through May 1979.

technical staff became so familiar with the model that they were able to predict the model's conclusions.

Modeling occurred in three phases: (1) from the NAPCTAC meeting in December 1977 to the time the NSPS was proposed in the *Federal Register* in September 1978; (2) September 1978 until December 1978 when additional model returns were published in the *Federal Register* and public hearings conducted; and (3) January 1978 after the public hearings to May 1979 when the standard became final.

Phase 1 modeling produced numbers in April and again in August. These were published in the September 1978 *Federal Register* when the NSPS was proposed. (See Appendix 2 for the conclusions published in September, 1978.) Effects were predicted for continuation of the 1971 NSPS, full control (0.2 lb. floor) and three partial scrubbing alternatives (0.5 lb., 0.7 lb., and 0.8 lb. floors).

For the April runs these changes were made over previous Teknekron modeling:

- 24-hour averaging instead of annual averaging;
- Alternative floors;
- Alternative ceilings; and
- Revised estimates for partial scrubbing costs.

The conclusions made available in April showed greater differences in costs and oil consumption between alternative standards than the first modeling work by Teknekron. For example, incremental annualized cost for full scrubbing (0.2 lb. floor) was estimated to be $2 billion in 1990 and for partial scrubbing (0.5 lb. floor) $1.3 billion. Utility oil consumption increased by 200,000 barrels a day under full scrubbing (0.2 lb. floor) over partial.

In summary, by raising the emission floor to allow more partial scrubbing, major impacts were:

- An increase in emissions from 1.5 million tons SO_2 on new plants fully scrubbed in 1990 to 2.1 million tons on partially scrubbed (0.5 lb. floor);

- Increased shipments of low-sulfur Western coal to the East. Western coal consumed in the East ranged from 299 million tons under full scrubbing to 346 million tons under partial with a 0.5 lb. floor to 429 million tons with a still higher 0.8 lb. floor;

- Lower utility oil consumption. Both partial and full scrubbing increase oil consumption over existing NSPS, but partial increases use less than full. Oil consumption is increased because new plants are more expensive and utilities will leave existing oil-burning plants on line longer;

- Lower annualized costs.

Major impacts of lowering the 1.2 lb. ceiling to cut emissions were:

- Increased Western coal shipments to the East, with a significant decline in Midwest coal production;
- Higher annualized costs;
- Sometimes a decrease in SO_2 emissions.

In general, the new modeling was showing more benefits for partial over full scrubbing than had been shown previously.

By the latter part of the period from April 1 through August, DOE began to become involved in the modeling process. EPA specified five alternative standards to be modeled, with varying floors, and with and without 3 days per month of exemptions from the 1.2 lb. emission ceiling. DOE specified two higher floors to permit less scrubbing than EPA. Also, DOE specified some sensitivity analyses to key inputs—electricity growth, oil prices, rail rate increases.

Impacts were projected to 1985, 1990, and 1995. These were compared to impacts of the current NSPS. Scrubbing costs were refined, the model refined, more attention paid to averaging periods, and assumptions changed. During this period, assumptions were changed to reflect lower growth rates for electricity demand, less growth in nuclear capacity, higher oil prices, and slightly higher coal transport costs and coal mining labor costs. (See Appendix 2 for the described changes in assumptions.)

By August, new modeling conclusions became available. This period produced some important conclusions. Annual emissions were considered to be "essentially equivalent" for full scrubbing and partial scrubbing with a 0.5 lb. floor.[60] Both reduce annual emissions of sulfur dioxide in 1990 by about 2.5 million tons or 12 percent. There are two reasons for this somewhat counterintuitive conclusion. First, the more stringent, full scrubbing standards make new plants subject to the new standard more expensive and, consequently, existing, dirtier plants are used longer. Existing plants average 1.0/lb. of SO_2/mill. Btu, whereas alternative NSPS have a maximum annual emission rate of 0.5 lb.[61] Thus, the additional emissions from old plants must be added to the emissions expected from fully scrubbed plants and this total about equals the emissions from partially scrubbed plants.

Second, overall emissions do not increase with partial scrubbing because partial scrubbing units which burn very low sulfur coal to comply with a higher floor without scrubbing actually have lower emissions than full scrubbing units which will burn high-sulfur coal, because it is cheaper.

While total emissions were predicted to be "essentially equivalent" for full and partial scrubbing, regional differences in emissions were significant. [In the East, emissions were about the same for a full (0.2 lb. floor) and partial (0.5 lb. floor)]. In the Midwest and West South Central emissions were reduced under partial as compared to full. But in the West, emissions increased with partial over full scrubbing. So while partial scrubbing was "essentially equivalent" in emissions nationally, in the West, current emissions would decline from the existing NSPS, but with partial decline by less than with full scrubbing. ICF judged the regional effects to be "credible and important,"[62] and gave these reaons for the regional emission effects:

> Emissions in the East do not change significantly between the alternative floors because the price of low-sulfur coal is very high in the East and hence little partial scrubbing is forecast to occur. The price differential between low- and high-sulfur coal more than offsets the scrubber cost savings of partial scrubbing low-sulfur coal rather than full scrubbing high-sulfur coal.
>
> Emissions in the Midwest and West South Central are forecast to go down in moving from the 0.2 floor to the 0.5 floor because partial scrubbing units with an emissions rate of 0.32 are substituting for (a) full scrubbing units on high sulfur coal with an emissions rate of 0.5; and (b) existing oil and coal plants with an emissions rate of about one pound.
>
> Emissions in the rest of the West increase in moving from 0.2 to 0.5 because the locally available coal is not high-sulfur so that partial scrubbing of low-sulfur coal with an emissions rate of 0.32 is generally substituting for full scrubbing medium-sulfur coal with an emissions rate of less than 0.32. Further, existing coal capacity is so small in the West that loads cannot be shifted from new to existing coal plants. Finally, the economics of coal versus oil generation in the West is such that the load on existing oil plants is unaffected by the cost savings of partial scrubbing.[63]

The model showed these other effects of partial vs. full scrubbing which also tended to favor partial scrubbing:

- More oil burned with full scrubbing than with partial—350,000 barrels a day for full (later shown to be 400,000) as compared to 200,000 barrels a day with partial;
- Full scrubbing costs are greater than partial—full scrubbing increases annualized costs by $2.3 billion in 1990, as compared to $2.0 billion for partial;

- Present value costs for full scrubbing are $26 billion, as compared with $23.2 billion for partial;
- National average electricity rate increases greater with full scrubbing—full scrubbing increases costs by 2 percent, partial by 1.5 percent.
- High-sulfur coal production increases with full scrubbing as compared to partial. With full scrubbing, low-sulfur coal in General Appalachia and Northern Great Plains drops. High-sulfur in Northern Appalachia and Midwest increases;
- With partial scrubbing, the value of low-sulfur coal increases somewhat.
- Use of high-sulfur coals increases, but by less than under full scrubbing.

In addition to showing benefits for partial scrubbing as compared to full, modeling from this period gave data on these other NSPS issues:

- The effect of no exemption to the ceiling was to prohibit the burning of high sulfur coal with a sulfur content greater than 1.6 lb. of sulfur/million Btu. Not only would high-sulfur regions be heavily impacted, but also costs of compliance would rise, since lower sulfur types are more expensive due to transportation costs. As a result, EPA began to think exemptions must be allowed;
- Sensitivity runs showed that reduced oil prices meant more oil consumption, less coal production, increased emissions, since 1 lb. SO_2 residual oil burned compared to about 0.5 lb. SO_2 in new coal-fired plants. As a consequence, runs showed that programs designed to reduce oil consumption will simultaneously reduce sulfur emissions. EPA and DOE could ally themselves along these lines;
- If rail rates did not escalate beyond the national inflation rate, oil consumption and emissions would be reduced, since more controlled new plants would be built, displacing older, dirtier plants. Also, consumption of Western coal in the East would increase substantially.

(See Appendix 2 for the modeling results from the August period.)

UARG'S MODEL

At the same time that EPA and DOE were jointly funding ICF model runs, UARG had hired its own modeling firm, the National Economic Research Associates (NERA) from New York. NERA used a model similar in design to ICF's, but was using different assumptions, and, of course, getting different results.

The outputs, similar to ICF's model, included predictions of NSPS impacts on:

- Coal production;
- Capacity additions;
- Scrubber capacity;
- Raw materials;
- Energy;
- Transportation;
- Operating and capital costs;
- Flue gas scrubbed;
- Sludge.

The final output of the model showed cost as a percentage of pollutant reduction. When inputs to the UARG and EPA models varied, results differed in aggregate numbers, but the two models were showing the same preference for variable percent reduction or partial scrubbing over full scrubbing. UARG delared:

> No EPA proposed regulation has been preceded by such an extensive macroeconomic analysis of potential costs and energy impacts by both EPA and DOE jointly (ICF) and by industry (NERA for UARG). Taking into account different input assumptions, the ICF and NERA results are remarkably consistent (parenthetical expressions mine).[64]

Then UARG went on to complain that full scrubbing advocates tried to discredit the modeling exercise,

> Proponents of the full scrubbing option first tried to downplay these economic and energy facts by changing the assumptions in the macroeconomic model analyses (world oil prices, electric utility growth rate, coal freight rates, etc.). When the results of the new assumptions were essentially consistent with the first results in failing to provide *any affirmative support* for the full scrubbing option and again supporting the partial scrubbing options, the proponents of full scrubbing sought to belittle this evidence by noneconomic subjective arguments:
>
> (a) Restating them in terms designed to make them look de minimus;
> (b) Raising doubts as to the usefulness of any macro-economic analyses in light of inherent uncertainties as to necessary assumptions;
> (c) Bringing in new Temple, Baker, & Sloan numbers at the last minute that are inconsistent with the ICF and NERA numbers and have never been discussed in the extensive interagency and industry interchange of economic information and joint discussions;
> (d) Resorting to the House Bill and Report to create a "statutory presumption" for a single percent reduction number apparently so strong as to override the express language of the Act and the conference Report.[65]

A great many numbers resulted from the modeling runs, and both full and partial scrubbing advocates used different ones to make their case. For example,full scrubbing people pointed to the small increase (2%) in the monthly electricity consumer bill that would result from full scrubbing and the increased emissions in the West from partial (100,000 tons more from 0.5 lb. floor by 1990). When talking about costs to industry they would point to the smaller figure of the annualized cost ($2.3 billion for full scrubbing), rather than to present value totals ($26 billion).

By contrast, partial scrubbing advocates would point to the multi-billion dollar price tag over the life of the plant that full scrubbing would mean in contrast to partial and the "essentially equivalent" emissions resulting from full and partial scrubbing nationally. The numbers were so complicated and voluminous—taking up thousands of report pages, changing almost weekly—that there was something in them for everyone. But the overall preference of the ICF and the utility industry's NERA models was slightly in favor of partial scrubbing. The numbers, however, did not paint a clear black and white picture. The effects of alternative standards were not vastly different and, indeed, varying the assumptions made bigger differences than the various alternatives using one set of assumptions. But the modelers were quick to point out that whereas changing assumptions altered the aggregate numbers, the relative preference for partial scrubbing remained.

When modeling results began to come in, showing advantages of partial scrubbing, Walter Barber would periodically brief Hawkins and Costle on the findings. The debates among technical staff as to modeling assumptions and their consequences and model types were not laid out to Costle, but were described in detail to Barber, Hawkins, Gamse, and Drayton. Costle had delegated the modeling issues to these officials for resolution.

However, Costle was extremely interested in the conclusions about partial vs. full scrubbing and directed that partial scrubbing be more specifically analyzed. OAQPS staff and Barber himself were beginning to believe the model, when it repeatedly showed significant cost and oil consumption differences among alternatives, but "essentially equivalent" emissions. Barber was looking more favorably to partial scrubbing after seeing the August modeling results, while Hawkins still pressed hard for full scrubbing.

One of the arguments that Hawkins and other full scrubbing proponents made was that full scrubbing would eliminate the need to negotiate "Best Available Control Technology" with each new power plant under PSD, and that this would save companies time and money in getting approval to construct new plants, which would offset some of the extra investment in control equipment.

However, OPE's David Shaver analyzed the cost associated with power plant delays due to BACT determination with partial scrubbing, and concluded, "that the potential for delay costs resulting from a 'partial scrubbing' standard are probably not significant and would not be a deterrent to industry support for the standard."[66]

5

THE SEPTEMBER *FEDERAL REGISTER* PROPOSAL

In this chapter, EPA decides to publish in the *Federal Register* the full scrubbing and partial scrubbing alternative NSPSs it is considering, stating that the administrator has not made up his mind as to the preferred option. How EPA came to this decision and the reactions to it are described.

THE ENVIRONMENTALISTS' GROWING CONCERN

During the summer of 1978, as this modeling was underway and modeling results and OPE were becoming more influential with OAQPS, the environmental groups saw EPA wavering on its previous commitment to full scrubbing. Delay, they figured, worked in favor of industry, as additional data and computer runs were pointing toward partial scrubbing and the industry had more time to lobby for its position. In May, Richard Ayres of the National Resources Defense Council, wrote Costle that the Clean Air Act only permits full scrubbing.[67] Also, environmentalists were looking to the NSPS decision as the crucial test of President Carter's commitment to environmental protection.

Compounding EPA's problems with the environmental community, a flap developed over the role of the White House inflation fighters in environmental and health regulations. The front page of the June 12 *New York Times* carried this headline: "Environmentalists Fear Cost Cutting: See Anti-inflation Actions Eroding Air, Water and Health Rules."[68]

Shortly after being named Mr. Carter's counsel on inflation, Robert

S. Strauss indicated that he saw reductions in the cost of pollution control as a major tool in fighting inflation.

The Regulatory Analysis Review Group (RARG) under Charles Schultze was seen also as undercutting EPA and other regulatory agencies. The *Times* story said:

> Mr. Strauss insisted, "I don't want to turn back environmental goals." Joking that "I didn't arrive in town under a load of watermelons," he added, "I couldn't push Doug Costle around if I wanted to."
>
> Mr. Costle, in an interview, tended to minimize the problem, saying, "This is not an argument about ends but means, and there is room for legitimate debate."
>
> Concerned environmentalists, however, question how legitimate the debate may be. Richard Ayres, an air quality expert from the Natural Resources Defense Council, said "I'm disturbed that Schultze and Strauss may be trying to do some things that are flatly illegal."
>
> He cited the attempt by Mr. Schultze and the Department of Energy to persuade the environmental agency to modify a regulation, now being formulated, on "performance standards" for new sources of pollution.
>
> *The Acid Test*
>
> "The new source standards are the acid test for the Carter Administration," Mr. Ayres said. "If they cave in on that, they forfeit their claim to the support of environmentalists' . . . "
>
> Moreover, almost everyone interviewed predicted that if the environmental community lost faith that Mr. Costle, who has enjoyed an excellent reputation, was acting independently there would probably be a flood of lawsuits by private organizations asking federal judges to order strict pollution controls.[69]

The environmental organizations' concern about White House economists undercutting environmental rules led them to conclude that these were "ex parte" communications—improper conversations that could influence decisions unfairly and in secret and therefore were legally forbidden.

On July 14, the Sierra Club went to court, asking that the August 7 date in the Clean Air Act for EPA to propose the NSPS be met. The court ordered EPA to propose the NSPS by September 12, and promulgate the final standards within 6 months. The utility industry agreed to these deadlines.[70]

THE WHITE HOUSE GETS INVOLVED

As the September deadline for proposing a NSPS grew near, the CEA and COWPS began to take active interest in the draft regulation and modeling analysis, pushing for partial scrubbing. DOE and the utility industry were both influential in piquing these White House units' interest in the NSPS, stressing the high projected costs and inflationary impact of the proposed rule. CEA and COWPS staff began to participate in the EPA/DOE joint task force meetings. William Nordhaus, a member of CEA, wrote EPA asking the agency to calculate benefits of alternative standards in terms of relative health impacts.[71] Within EPA, Barber decided that there was no time and developed methodology to do benefits analysis.[72]

The analysis EPA was conducting is called *cost-effectiveness analysis*, in which the cost of reducing alternative levels of emissions was projected. Cost-benefit analysis would go two steps beyond that, estimating the impact of alternative standards on ambient air quality conditions around the country and, further, the effect of those air pollution levels on health, human welfare, and visibility. Such projections would require complicated and expensive dispersion modeling, based on estimates where power plants would be located. Dose-response studies would also have been necessary, to predict the impacts of resulting pollution levels on human beings. Visibility studies would be required since little is known about the impacts of sulfur and particulate emissions on views.

By August, DOE's O'Leary urged a variable percentage (partial control) standard, ranging from 33% to 80% SO^2 removal with an emissions floor of 0.8 lb./mill. Btu of SO_2.[73]

As the rule was being written by EPA to go to the *Federal Register* interest groups and others were taking these positions:

For Full Scrubbing

- Environmentalists;
- Environmental committee staff on Capitol Hill in the Senate Public Works Committee and House Interstate and Foreign Commerce Committee;
- Assistant Administrator Hawkins, some OAQPS staff;
- Officials in the East, particularly Appalachia, who wanted full scrubbing to encourage use of local high-sulfur coal.

For Partial Scrubbing

- Utility industry;

- Business groups;
- Council on Economic Advisors, Council on Wage and Price Stability, who saw this as less costly, less inflationary;
- Department of Energy, concerned about oil impacts and cost to utilities;
- EPA's Assistant Administrator Drayton, his deputy Roy Gamse, and OPE;
- Energy committee on Capitol Hill;
- Western coal interests, since partial scrubbing will increase demand for low sulfur western coal.

While Costle was no longer sure that full scrubbing was the best alternative NSPS, he did believe that the Clean Air Act included a statutory presumption in favor of full scrubbing. Based on this thinking, he was preparing to include in the *Federal Register* full and partial scrubbing alternatives, but to state a preference for full scrubbing. Costle described his thinking in an August memo to Stuart Eizenstat, advisor to the president:

> There is insufficient time to settle the questions on this standard being discussed both inside EPA and in the administration generally, much less to both settle the questions and write the solutions into acceptable regulatory form. Nor is there any legal need to resolve them; the purpose of a proposed standard is to invite comment on alternatives and help me make up my mind.
>
> Our proposal will state explicitly that the Agency has not made up its mind on the final standard, and will contain a full discussion of each of the main alternatives and a call for comment on them.
>
> In normal rule-making, we could stop there. However, the Clean Air Act provides that new source performance standards apply from the date they are proposed, not, as it is more usual, from when they are promulgated. We therefore think we must designate one option as preferred for the guidance of sources that begin construction between proposal and promulgation.
>
> We plan to designate full scrubbing for all plants as that option, not because we necessarily favor it as the final standard, but because it is the only choice that we can make consistent with the basic position that we haven't made up our minds yet. We will rest that choice on two grounds:
>
> (a) I believe everyone, including DOE, concedes that the statute contains a presumption in favor of full scrubbing. That presumption might be overcome by detailed analysis of the factual impacts of various standards. However, since we haven't com-

pleted our analysis and reached our conclusions yet, we cannot say the presumption has been overcome; [and]

(b) It will be easier for power plants that start construction between proposal and promulgation to scale down to partial scrubbing if we proposed full scrubbing and then changed our minds than it would be for them to scale up if we do the reverse. This consideration of relative burdens in itself supports proposing full scrubbing.[74]

Referring to environmentalists' demands that if CEA, COWPS, and other White House staff intervene in the rule-making proceedings it be on the public record (no ex parte contacts), Costle continued:

Mid-March is a very tight deadline for issuing a final standard considering the scope of the issues. To issue a good standard by then we need the cooperation of all interested agencies. To be explicit, we need their comments on the issues in as much detail as possible, and we need them furnished *for the public record during the public comment period.*

Politically, I do not think we could justify promulgating a standard of this importance on the basis of comments or analysis that the interested public had not been able to examine. Such a course might well also cause a court to hold that EPA had failed in its obligation to follow open rulemaking procedures as required by the Clean Air Act. There are already cases to this effect in the D. C. Circuit which reviews all new source performance standards.

By requiring that the public record be complete as to other agencies' views and analysis, subsequent internal debate within the administration following close of the public comment period and before a decision in March can be candid and uninhibited. To the extent that that discussion involved issues or analysis not fully presented in the public hearing record, I cannot legally allow them to influence my decision.

I personally feel *very* strongly about this and believe that the credibility of this administration's pledge to do the public's business in public is very much at stake. As you are aware, the President is receiving increasing criticism concerning the manner of White House (specifically CEA and COWPS) intervention in regulatory actions. Much of the concern being expressed is unwarranted. You should not for a moment, however, underestimate the seriousness of the concern being expressed or the potential for irreparable damage to this administration's reputation.

I am fully desirous of complete, thorough, and responsible advice from the CEA, DOE, or any other agency of the federal government

> before I reach a final decision on this standard. However, it is critical that the *manner* in which that advice is both given and received be responsible and open.[75]

In August, Costle met with presidential advisor Stuart Eizenstat, DOE's Schlesinger, and EPA's general counsel, Jody Bernstein, to brief the administration officials on what EPA would propose in September.

Costle indicated the September proposal would list several NSPS alternatives, saying he had not made up his mind, but because of the legalities would cite a preference for full scrubbing. He would also indicate that there was a "statutory presumption" in the Clean Air Act for full scrubbing, but that this presumption could be overcome by other considerations. The sense of the proposal would be a preference for full scrubbing.

Schlesinger was able to persuade EPA to change this language to remove the bias for full scrubbing and instead present alternatives with no stated preference. Schlesinger urged that the "statutory presumption" statement be dropped, that this would be prejudging the issue. At another meeting in September, prior to publication of the proposed standard and attended by Costle, Hawkins, Schlesinger, and Eizenstat, Eizenstat sided with Schlesinger in what Hawkins described as an "acrimonious" meeting.[76] Costle and Hawkins agreed to omit reference to "statutory presumption." The overall effect was to make the proposal more neutral in tone from the previous language. While technically the package was unchanged with NSPS alternatives presented, politically it was altered with government support for full scrubbing toned down.

After this meeting, Costle and Hawkins realized that there could be a major administration battle over promulgation of the final standard if not handled correctly and they began to plan their course of action for White House review. They knew it would be even more important to resolve technical debates and leave discussion at the policy level. Modeling and other technical analyses would be stepped up and efforts made to involve all pertinent government actors in agreement on technical facts through the EPA/DOE Joint Task Force.

THE SEPTEMBER PROPOSAL

The NSPS proposal in the *Federal Register* offered several alternatives for comment, saying the "Administrator has not made a decision on which of the alternatives should be adopted."[77]

Full scrubbing is presented first because "the Clean Air Act provides that new source performance standards apply from the date they are proposed and it would be easier for power plants that start construction during the proposal period to scale down to partial control than to scale up to full control should the final standard differ from the proposal."[78]

This option was basically as proposed to NAPCTAC. The parts revised from early EPA draft are underlined below:

SO_2:

- 1.2 lb. mill. Btu emission ceiling, except for 3 days per month.
- 85 percent* reduction of uncontrolled sulfur emissions, averaged every 24 hours with 3 days exemptions per month when 75 percent reduction would apply.
- A 0.2 lb./million Btu emission floor. Compliance with the emission floor would constitute compliance with the percentage reduction. (This standard as a result would require more than 85 percent reduction for coals with uncontrolled SO_2 content greater than 5.3 to 8.0 lb. For coals of 3.7 lb. to 5.3 lb., the 85 percent removal requirement would be the binding figure; and for coals below 3.7 lb. sulfur the emission floor becomes the binding constraint and less than 85 percent removal is required—about 80 percent of 1 lb. sulfur coals. Since scrubbing would be necessary to achieve 80 percent removal, this standard meant full scrubbing.)
- Credit would be allowed for precombustion fuel cleaning. Sulfur removed in the pulverizer or in bottom ash and fly ash would be credited toward the percent reduction requirement;
- Low-sulfur anthracite coal combustion would be covered;
- So as not to discourage emerging front end control technology which can be capital intensive, the administrator, in consultation with DOE, would issue commercial demonstration permits for the first three full-scale demonstration facilities of solvent refined coal, fluidized bed combustion (atmospheric), fluidized bed combustion (pressurized), and coal liquifaction. Under such permits 80 percent SO_2 removal, averaged daily, or 0.70 lb. NO_X emission limit for liquid fuel derived from bituminous coal would be required.

The particulate and NO_X standard were the same as those proposed to NAPCTAC. The DOE 33 to 85 percent sliding scale standard was presented also for public comment in the *Federal Register:*

- An 85% reduction of potential SO_2 emissions during each calendar month. Bypassing allowed as long as the percent reduction is met;

*The draft NSPS presented to NAPCTAC was 90 percent reduction but this had been calculated on an annual averaging, which was subsequently converted to 85 percent reduction, averaged every 24 hours.

- A 0.80 lb./mill. Btu SO_2 emission floor not to be exceeded during any 24-hour period. A sliding scale percent reduction required up to 85 percent on high sulfur coals. Only minimum percent reduction enforced for 24-hour periods, when SO_2 would be 0.8 lb. or less. Therefore, emissions must be greater than 0.8 lb. and reduction less than 85 percent to constitute a violation of percent reduction;
- A minimum of 33 percent reduction of potential SO_2 emissions to ensure that even very low sulfur coals (below 0.8 lb.) could not be used untreated.

This DOE proposed standard came to be called the "sliding scale" standard, because percent reductions slide between 33 and 85%, depending on the sulfur content of the coal. One can also substitute the words "variable" standard, which was the terminology used by the utility industry, or "partial scrubbing."

In addition to the DOE proposal, UARG suggested a standard based on a 20 to 85 percent sliding scale which was also published for comment:

- Ceiling of 1.2 lb. and a required percent reduction ranging between 85 percent removal on a coal with uncontrolled emissions of 8 lb. of SO_2 to 20 percent removal on coals with uncontrolled emissions of 1 lb. or less;
- Compliance with these sulfur dioxide standards would be averaged on a 30-day basis;
- Industry also asked for consideration of 1.5 lb. emission ceiling.

The *Federal Register* proposal carried the results of the impact analyses completed in April and in August. (These appear in Appendix 2.)

ENVIRONMENTALISTS' COMMENT ON SEPTEMBER PROPOSAL

The Natural Resources Defense Council (NRDC) and the Environmental Defense Fund (EDF) jointly commented on EPA's proposal, opposing both EPA and DOE's proposals as inadequate and offering this alternative standard:

- Uniform percentage reduction of SO_2 of 95 percent, averaged daily;
- Ceiling or maximum emissions of 0.8 lb./million Btu; and
- Coal washing credits allowed.

The environmentalists' position was that the Clean Air Act only permits uniform (full) scrubbing of all coals. Furthermore, the health and environmental risks from a variable (partial scrubbing) standard were substantial.

The "ceiling should be set so as to compel plants that choose to use higher sulfur fuels to achieve more than the minimum emission reduction."[79]

EPA's floor of 0.2 lb./mill. Btu allowed a few plants to scrub less than 85 percent (80 percent on 1 lb. sulfur coals), which the environmentalists said amounts to an illegal variable scale, which the Act does not allow.

The 95 percent removal figure the groups suggested came from the Japanese experience using scrubbers on low- and medium-sulfur coals. The 95 percent removal of SO_2 was achievable with scrubbers alone on low-sulfur coals and on high-sulfur coals with supplementary coal cleaning, the environmentalists said. The environmental groups cited conclusions of two American task forces that had visited Japan to evaluate scrubber technology. First, the Jackson Task Force cited earlier reported 96 percent removal rates. Additionally, a team of pollution control experts from the California Air Resources Board (CARB) and several California utilities visited Japan and reported:

> All of the flue gas desulfurization (FGD) systems were operating with efficiencies of over 90 percent, with an average of about 94 percent. If the addition of sulfur dioxide to the stack gases from the stack gas reheaters were excluded in the efficiency calculations, efficiencies ran as high as 99.4 percent."[80]

As a result of their visit, the CARB concluded that scrubbers are easily capable of achieving 95 percent SO_2 removal on a continuous basis with a very high degree of reliability.

In its comments on EPA's proposed NSPS, CARB called for a nationally uniform standard of 95 percent reduction of SO_2 with allowances for coal washing. CARB indicated that this could be met continuously, meaning by hour or by the minute. Averaging time should be no longer than 1 or 2 hours, with no exemptions. CARB comments to EPA included vendor guarantees to achieve 95 percent reduction and a guarantee by Pacific Gas and Electric Company to remove 95 percent of SO_2, on an hourly or instantaneous basis.

The environmental groups cited this CARB position to support their case for a uniform 95 percent SO_2 removal standard.

These groups argued that EPA's proposed 85 percent SO_2 removal requirement, averaged daily, did not go beyond the status quo, and the Act requires EPA to set NSPS to be technology forcing. Also, EPA's allowance of 3 days of exemptions is not needed, they said, because scrubbers will operate reliably if properly operated and maintained. By contrast, DOE contended that "there is no utility scrubber system in the world where it has been demonstrated that daily emission reductions for high-sulfur coal can be maintained at or above 85 percent for a sustained period of time. Successful Japanese experience has been with low- and medium-sulfur coals."[81]

As to the modeling, NRDC/EDF had no kind words, believing that faulty modeling assumptions biased the conclusions toward partial scrubbing. The conclusions that emissions would be "essentially equivalent" for partial and full scrubbing, but that oil consumption would increase markedly under full scrubbing resulted, they said, from an erroneous assumed price of oil and the assumption of "substitutability."

> They exist because of the assumptions the modelers have made about the price of oil in the future and because they have assumed that the utilities will choose between building new coal-fired power plants, on the one hand, and stretching out the lives of old, poorly controlled oil and coal-fired power plants or building new oil turbines, on the other, entirely on the basis of the assumed cost difference between these alternatives at the time they make their decision. Implicit in this assumption is also the assumption that the utilities will be entirely free to substitute these alternatives for building coal-fired power plants conforming to the NSPS. Given these assumptions, the model predicts that the small difference in cost between uncontrolled and controlled coal-fired power plants will be sufficient to induce the utilities to rely far more on old plants and new oil-fired turbines.
>
> So long as the model assumes oil prices within the range chosen by the modelers and adopts the hypothesis that old oil- and coal-fired capacity can be freely substituted for new coal-fired capacity, it will minimize the projected advantage in reducing emissions of full scrubbing over partial scrubbing, and minimize the apparent difference in oil consumption associated with the two. . . . If the assumed oil price is too low, or if substitutions will be prevented by economic risk-averse behavior by utilities or by non-economic factors, or could be prevented by conscious government intervention, then the modelers have grossly understated the difference in environmental benefits between full and partial scrubbing alternatives."[82]

Environmental spokesmen also contended that the capability was lacking to conduct "cost/benefit analyses as a basis for deciding a proper NSPS" and that "cost-effectiveness" is not a "sound basis for deciding what level of control the NSPS should require . . . since it generates figures that, even when accurate, have no context whatsoever."[83] (For additional environmentalist reaction to the September proposed NSPS, see Appendix 3.)

DOE staffer Dan Badger subsequently responded to these environmentalists' criticisms of the modeling assumptions. Changing such modeling assumptions as electricity growth rates and coal transportation costs apply equally under either NSPS alternative, Badger wrote, so that "although the *absolute* levels of costs and emissions that would result . . . are quite uncertain, the *differences* between the alternatives can be projected with much greater certainty."[84]

The one assumption that would change the relative cost/effectiveness of full and partial scrubbing is cost of scrubbers, but Badger continued,

> This cost differential (between full and partial scrubbing) is not one of the parameters over which there is any substantial uncertainty. Partial scrubbers are made up of components of full scrubbers. Thus, while there may be uncertainty over the absolute cost of either one, the relative costs can be estimated with much higher confidence. This is because the major factors that create uncertainty such as cost escalation rates, apply equally to full and partial scrubbers (parenthetical expression mine).[85]
>
> (The environmentalists) are also correct in noting that the relative cost-effectiveness of the alternatives is sensitive to the oil price assumption. But they incorrectly predict that higher oil prices make full control relatively more cost effective. In fact, the opposite is true. Although higher oil prices would reduce the extent to which full control would "stretch out" use of oil-fired plants, the higher price of oil consumed in units that remain in operation longer would offset this and *raise* the cost of the full control option.[86]

DOE'S RESPONSE TO SEPTEMBER PROPOSAL

DOE, in its formal comments on EPA's September NSPS proposal, argued that its preferred sliding scale (partial scrubbing) standard was more cost-effective than full scrubbing and would reduce oil consumption. DOE called for phasing in a 93.5 percent monthly average reduction in two periods.

> The modifications we propose would achieve substantial cost and energy savings, without sacrificing any of the environmental benefits that would result from EPA's proposals. Indeed, our option, the "sliding scale" proposed for the SO_2 emissions and sulfate concentrations, because it would reduce utilization of existing power plants, and lower the average sulfur content of coals used in new plants. These findings indicate that the sliding scale would fulfill environmental goals, while supporting more effectively the president's recent initia-

Table 5.1: Summary of DOE Recommendations for Utility NSPS

1. *Sulfur Dioxide Standard:*

	NSPS Revision Cycle[a]	
	1979	*1983*
Maximum percent removal requirement	90% averaged monthly, including washing credit	93.5% averaged[b] monthly, including washing credit
Minimum percent removal requirement	33% averaged monthly	33% averaged monthly
Maximum emission rate (emission ceiling)	Monthly average equivalent of 0.8 lb. SO_2/mill. Btu annual average	Monthly average[b] equivalent of 0.55 lb. SO_2/mill. Btu annual average
Maximum control level (emission floor)	Monthly average equivalent of 0.55 lb. SO_2/mill. Btu annual average	Monthly average[b] equivalent of 0.55 lb. SO_2/mill. Btu annual average

2. *Particulate Standard:*
   Maximum emission rate of .05–.08 lbs per million Btu, revision to 0.03 lb. following assessment of 1 year's data on currently operating utility baghouse installations.

3. *NO_x Standard:*
   Support the September 19 *Federal Register* proposal.

4. *Emerging Technologies:*
   Support the September 19 *Federal Register* proposal.

[a]The 1983 standard would come into effect automatically unless the administrator makes a negative finding after reviewing the data on FGD performance on high-sulfur coal.

[b]The standard for the second cycle could be implemented either by raising the maximum removal requirement to 93.5% or lowering the maximum emission rate to the monthly equivalent of 0.55 lb. annually. If the latter approach is chosen, the minimum removal requirement and the ceiling together would achieve the same result as the maximum removal requirement and the floor. These could, therefore, be eliminated.

tives to control inflation and to achieve greater cost-effectiveness in government regulations.[87]

Table 5-1 summarizes DOE's recommendation.

UARG'S COMMENTS ON SEPTEMBER PROPOSAL

The utility group:

- First favored UARG's sliding scale of 20 to 85 percent reduction of SO_2 and second, DOE's sliding scale of 33 to 85 percent removal over EPA's uniform scrubbing. The sliding scales were less costly and more cost-effective. National macroeconomic and energy effects of DOE and UARG proposals are roughly equivalent. "UARG is not contesting revision of NSPS so as to require scrubbers (or equivalent new 'front-end' technologies) on all new coal-fired plants";[88]

Contended that

- The sliding scale is permitted by the Clean Air Act; indeed, that there is no "statutory presumption" in favor of uniform percentage reduction of SO_2;
- sliding scale encourages use of lower-sulfur coal;
- Sliding scale produced less sludge and is therefore more desirable on "non-air quality environmental effects";
- Sliding scale results in less oil consumption than EPA's alternative, also less lime production required under sliding scale with its energy penalty;
- Sliding scale encourages innovative technology—front-end technologies such as solvent refined coal, dry scrubbers, fluidized bed combustion;
- Employment in high-sulfur coal regions will not be impacted significantly by the sliding scale, since the model shows that growth in demand occurs in all coal regions under all alternatives (although the model showed that growth will be less under sliding scale than full control);
- Particulate standard of 0.03 lb./million Btu requires electrostatic precipitators or baghouses that are not "adequately demonstrated" at this removal rate.

RARG'S REPORT ON SEPTEMBER PROPOSAL

On January 15, 1979 RARG reported on the NSPS, supporting the DOE sliding scale (partial) standard as more cost effective and less oil consumptive. Since September the COWPS and CEA authors of this report had been taking part in EPA/DOE task force meetings about modeling techniques and assumptions. The report relied on the results of EPA/DOE technical analysis to date and in fact EPA's OPE participated in development of the report. Relying on EPA modeling, which it called "one of the most thorough analytical efforts ever undertaken by a regulatory agency in preparation of a final rule,"[89] the RARG document concluded "the full scrubbing option would impose greater incremental and total costs than the sliding scale. Though total nationwide emissions are lower under full control (3 to 4 percent) than under the sliding scales, the emissions differences (and therefore differences in health impacts and other benefits) shrink substantially when emissions are weighted by regional population densities."[90]

As EPA had expected, Nordhaus of CEA was able to include in the report a criticism of EPA for failing to perform benefits analysis. The report went on to present its own human exposure index, a type of benefits analysis, which favored the sliding scale. This index showed that the sliding scale reduced more emissions in the Midwest and parts of the East where a greater population would be exposed to pollution. In the West where emissions increase under a sliding scale, human exposure is less. This benefits calculation did not attempt to predict impacts on actual air quality in specific regions nor to relate those air qualities to health effects in specific numbers of individuals. The RARG report had very little impact on the fast moving events to complete the NSPS.

6

RESOLVING THE ISSUES

In this chapter EPA leaders select the variable scrubbing standard over uniform scrubbing as well as resolving the lesser issues of monitoring requirements, averaging times, and treatment of anthracite coal. The second phase of modeling is concluded, raising the issue of the emissions ceiling, and the National Coal Association and Senator Robert Byrd become actively involved. After the EPA administrator approves a variable standard and permits use of dry scrubbing, a strategy for securing White House approval of the NSPS is devised and implemented.

MONITORING/AVERAGING TIMES

After receiving public comment on its September proposal in the *Federal Register*, EPA, working through its various divisions and along with DOE representatives, drew conclusions on the issues of monitoring, averaging times and anthracite coal.

Continuous monitors would be required to determine compliance, but the proposed 24-hour averaging period, with 3 days of exemptions a month, was changed to be a rolling 30-day average that obviated the need for exemptions. In other words, the average would be composed over 30 days, in each day a new day of data would be added and the oldest day of data dropped.

The 24-hour averaging period had been initially included because the EPA enforcement division needed daily numbers to assess the daily penalities for noncompliance, as required by the 1977 Clean Air Act Amendments.

However, as public comment indicated and additional EPA investigations verified, there was considerable uncertainty over causes and extent of short-term emission variations and what can be done to control variability. For example, both the sulfur content of the coal may vary more than the utility company intended and the control equipment may not function as uniformly or as efficiently as planned. Furthermore, the previous system of accounting for variability of performance had been to allow 3 days of exemptions a month, but this was thought to be a cumbersome provision from both an engineering and enforcement standpoint. Finally, there was no technical reason why the 24-hour period was needed to represent "best technology."

Opposition to the 24-hour period came from industry, DOE, and EPA's OPE. The proposed short averaging period was difficult to achieve, or at least expensive, and would push utilities to burn lower-sulfur coal for protection against a violation, they said.

Consequently, the 24-hour period was changed to a rolling 30-day average. The rolling provision was added to satisfy enforcement needs, since daily numbers are calculated in this system, and can be used to assess daily penalities.

Changing the averaging time required that new figures be set for percent reduction, ceilings, and floors. For example, a 75 percent reduction of SO_2, averaged every 24 hours with no exemptions, was equivalent to 85 percent reduction averaged every 24 hours with three exemptions, which was equivalent to 90 percent reduction averaged every 30 days. Table 3-1 shows that when the averaging period is lengthened, the emission ceilings number goes down, with the same emission effect.

These conversions were not only complicated statistical issues but became political issues as well, since outside parties that did not fully understand the change of averaging period thought that EPA was changing the stringency of the standard, when, in fact, only the averaging period was changing.

The second monitoring consideration was how much of the time the monitor had to be in operation to provide statistically significant data to calculate the average.

EPA's enforcement division was concerned that the data be sufficient to stand up as valid in court as demonstrating compliance or noncompliance and that procedures not allow a source to circumvent a valid reporting system in such a way as to disguise violations. Consequently, the enforcement division regularly reviewed drafts of the OAQPS/OPE task force and drafted language for inclusion.

The September proposal only allowed 1 hour per day for a continuous monitor to be out of operation, which would require installation of backup continuous monitor systems or reliance on manual testing when a continuous monitor system malfunctioned. The OAQPS staff working with the enforcement staff recommended a change so that more time would be allowed for monitor malfunction, yet be statistically satisfactory to demonstrate compliance. Under the recommendation the owner or operator would be allowed to have a continuous monitor system out of operation for 6 hours in any 24-hour period and 8 days in any 30-day period of boiler operation. This would eliminate the need for redundant monitors. The final standard includes a series of statistical tests of significance.

The minimum data required to be representative of a 30-day rolling average is 22 days with 18 hours each day. This conclusion was a compromise between EPA's enforcement needs and a desire not to be unreasonable with utilities.

ANTHRACITE COAL EXEMPTED

Most of the public comment on EPA's September proposal opposed inclusion of anthracite coal in the NSPS provisions. Anthracite coal, found exclusively in eastern Pennsylvania, is relatively low in sulfur, but it is 20 to 50% more expensive to mine and to burn than high-sulfur bituminous varieties. Since 1940, demand had greatly decreased and the anthracite region became severely depressed. Only if this coal were exempt from the scrubber requirements, it was said, could it become competitive again with the dirtier varieties in the area. Furthermore, it was argued, reopening anthracite mines would result in improvement of acid mine water conditions, eliminate old mining scars on the topography, eradicate dangerous fires in deep mines and culm banks, and create new jobs in the depressed area.

In consideration of these facts, EPA staff recommended and Costle agreed that the environmental and economic situation was unique and that anthracite coal be exempted from the percentage reduction requirement. No FGD systems would be needed and anthracite burners would be subject only to the 1.2 lb./mill. Btu 30-day rolling average emission limit, which would easily be met without end of system control equipment. The same NO_X and particulate matter standards that apply to bituminous coals apply to anthracite.

PHASE 2 MODELING: THE CEILING ISSUE EMERGES

The second phase of modeling occurred between September 1978 when the NSPS proposal was carried in the *Federal Register* and

December 1978 when a second *Federal Register* notice carried additional modeling results. During this period the interagency task force was active with CEA and COWPS staff taking part, as well as DOE and EPA. The main purpose of modeling in this period was to analyze the alternative NSPS with the new modeling assumptions agreed to by the interagency work group. The intent was to discern the most cost-effective standard.

Some model assumptions were changed. Oil prices used in the August analyses were considered too high, and so two lowered sets of oil prices were included. This round of analyses included coal washing credits for the first time. Flue gas desulfurization costs were changed slightly, including an error corrected in partial scrubbing costs and other lesser changes were incorporated. These changes resulted in relatively higher partial scrubbing costs when compared to full scrubbing. (See Appendix 4 for conclusions from this modeling period which were published in the December *Federal Register*.)

Results of modeling in this period showed the emissions ceiling or maximum emissions permitted to be a key issue.

Ceilings became important for two reasons. First, washing credits had been omitted from modeling runs and when finally included showed sizeable emissions increases. Washing would reduce scrubbing requirements of the standard from 90 to 85 percent. The cost savings of dropping from a 90 to 85 percent scrubbing requirement are small and the cost of washing is relatively inexpensive per ton of SO_2 removed. Therefore, the NSPS working group sought a low emission ceiling that would require washing and 90 percent scrubbing, substantially reducing emissions in the East and Midwest at a relatively low cost. Since washing is in widespread use, the EPA staff thought that Eastern coal production would not be seriously hurt. ICF modeling confirmed these conclusions. While the Clean Air Act made credits for precombustion coal cleaning optional, once EPA had proposed such credits, staff felt that withdrawing them would be virtually impossible, politically.

However, the main reason the ceiling became important was that previous modelings had made it clear to EPA and DOE analysts that the most cost-effective standard was a low ceiling with no percent reduction requirement. Such a standard would create incentives for utilities to purchase low sulfur coal and reduce expenditures for scrubbers. However, the Clean Air Act would not allow a standard without a percent reduction requirement. The next most cost-effective standard was shown to be a sliding scale or variable standard and a low ceiling. DOE and EPA wanted to model several alternative ceilings to determine which was most cost-effective.

Standards with either no percentage reduction or a sliding scale were more cost-effective than full scrubbing because of the "Midwest anomoly."

In areas of the Midwest, both low- and high-sulfur coal can be bought at a reasonable cost, but the delivered cost of high-sulfur coal is somewhat less. The removal requirement, therefore, becomes the decisive factor in which coal is economically preferred. If full control is required regardless of the type of coal used, high-sulfur coal will be bought. If partial control is allowed for low-sulfur coal, low-sulfur coal will be burned, if the savings in control costs are enough to offset the increase in fuel costs.

Emissions from partial scrubbing low-sulfur coal are less than emissions from scrubbing on higher sulfur coal. If the choices are 90 percent control on 8 lb. sulfur coal or 33 percent control on 1 lb. coal, the emission rate under partial scrubbing would be lower.

While lowering the ceiling would reduce nationwide SO_2 emissions, EPA knew that it would also reduce demand for high-sulfur coals, such as those from the Midwest and Appalachia, just as partial scrubbing would decrease demand for low-sulfur coals.

A very low ceiling would make the highest sulfur varieties unburnable in utility boilers because even with effective scrubbers sufficient SO_2 could not be removed in utility boilers. This would exaccerbate unemployment problems in those already economically depressed regions of high-sulfur coal mining and create substantial political problems from the Midwest and Eastern coal industry, the UMW whose members are mostly in these fields and their political representatives. Furthermore, the Clean Air Act directed EPA to encourage the use of locally available coal, and any very low ceiling would not do this. EPA wanted the model to answer how much impact would occur on high sulfur coal reserves from alternative ceilings.

DOE's Dan Badger, in consultation with ICF, estimated that a 0.55 lb. annual average emission ceiling would be the most cost-effective. The number was put to the model which confirmed that this was so. Consequently, DOE proposed the 0.55 lb. annual average emission ceiling in response to the September *Federal Register* proposal. (See Table 5-1.)

While Badger and other technical staff knew such a low ceiling would increase demand for low-sulfur coal, it was not known, at first, that it would have a very substantial impact on high-sulfur coal markets.

DOE suggested that the 0.55 lb./mill. Btu emission ceiling accompany a variable percentage reduction figure of 33 to 90 percent. EPA suggested ceilings be modeled of 0.8 lb./mill. Btu annual average (which it said at that time was equivalent to the 1.2 lb./mill. Btu daily average contained in the September proposal).

Both uniform and variable percent reduction alternatives were modeled. Alternatives specified by the Utility Air Regulatory Group and the Natural Resources Defense Council were put to the model, but since the modeling assumptions offered by NRDC did not conform to the government assumptions, the results could not be compared.

On December 8 EPA published the results of this modeling period, including these alternative ceiling runs in the *Federal Register*, to advise the public of the conclusions. (See Appendix 4 for conclusions on this modeling.) Public hearings on the findings were held on December 12 and 13. Effects of alternative standards on emissions, economics, energy production, and consumption were presented. National coal use was projected in terms of production and consumption by geographic region. The amount of Western coal shipped to the Midwest and East was also estimated. In addition, utility consumption of oil and gas was analyzed.

The published results demonstrated in a clear way to the public that EPA technical staff was seriously considering a substantially lower emission ceiling than was previously known. The published figures further showed that the very low ceilings would preclude the burning of some higher-sulfur varieties of coal found in the Midwest and Appalachia.

Administrator Costle, throughout the rule-writing process, thought the standard should not cause major disruptions in high-sulfur coal markets. He did not want EPA actions to be responsible for substantial unemployment in the coal fields and the consequent political problems that would bring EPA. Labor unions had traditionally been an EPA ally, and Costle wanted the UMW to be with EPA when it came time for final rule approvals and not aligned against the Agency on the NSPS. Furthermore, the Clean Air Act was written to protect "locally available coal," as a result of the alliance between the UMW, high-sulfur coal employers, and environmentalists. Consequently, Costle asked Barber to make more detailed analyses of impacts on coal reserves in the East, Midwest, and northern Appalachia, the highest-sulfur coal regions. EPA realized that the ICF model, by necessity, contained a number of simplifying assumptions on coal characteristics and that the model placed no value on the social costs of economic dislocations. Thus, the ICF macroeconomic analysis was supplemented by an EPA staff microanalysis of coal market impacts. The analysis, begun February 1979, was based on Bureau of Mines reports which provided seam-by-seam data on the sulfur content of coal reserves. It showed which coals would require more than 90 percent scrubbing of washed coal in order to meet designated ceilings. The assessment was made using Battelle's Reserve Processing Assessment Model, developed by EPA's Office of Research and Development.

Data from the microanalysis showed, for example, that with 85 percent scrubbing and an emission ceiling of 0.55 lb. SO_2/million Btu calculated on an annual average, 75 percent of coal reserves in Illinois, Indiana and western Kentucky could not be burned; 35 percent could not be burned in northern Appalachia (Pennsylvania, Ohio, and West Virginia); and 30 percent could not be burned nationally. As scrubbing efficiencies increased from 85 percent to 90 percent and 95 percent and more advanced coal preparation techniques were employed, considerably more high-sulfur coal could be burned. The analysis showed that with a 0.55 lb. SO_2/mill. Btu emission limit (annual average) more than 90 percent scrubbing would be required on between 5 and 10 percent of northern Appalachian reserves and on 12 to 25 percent of the Eastern Midwestern reserves. At a 0.8 lb. SO_2/mill. Btu ceiling (annual average which is considered 1.2 lb. daily), less than 5 percent of the reserves in each of these regions would require greater than 90 percent scrubbing.

Table 6-1 presents the percentages of burnable coal in the Midwest and northern Appalachia under a range of alternative NSPS ceilings. Impacts are shown with three scrubbing scenarios—85, 90, and 95 percent.

Reinforcing the view that EPA was getting serious about a low emissions ceilings, several memoranda were circulated within EPA assuming a 0.55 lb. (annual average) ceiling. A January paper written by John Haines stated, "We have focused on the alternatives that were modeled with a 0.55 lb./mill. Btu annual emission ceiling. Three alternatives appear to have the most merit: one full scrubbing, one partial scrubbing, and one regional option."[91]

Then David Tundermann of OPE wrote Bill Drayton that the "current recommended alternatives are:

Full scrubbing: 90% removal (annual average), 0.55 lb. SO_2/million Btu emission ceiling,

Partial scrubbing: 33% minimum removal (annual average), 0.55 lbs. SO_2/million Btu emission ceiling,

Regional standard: full scrubbing in the West (11 states), partial scrubbing elsewhere."[92]

On March 14, Costle and his staff briefed key administration officials about these three basic scrubbing alternatives under study—

Table 6.1: Burnable Coal Reserves Under Alternative SO_2 Emission Limits
(Percent of Btu)

Coal Region	Coal Reserves (10^{15} Btu)	1975 Annual Production (10^{15} Btu)	Emission limit, lb. SO_2/mill. Btu annual average 0.55	0.65	0.70	0.80	1.0*
			(Percent of Btu)				
North Appalachia (PA, OH, WV) (%)							
85 FGD	1730 (50)	6.0	65 (12)	75 (37)	85	90 (63)	95 (97)
90 FGD			90 (63)	95 (90)	>95	>95 (99)	>95 (99)
95 FGD			>95	>95	>95	>95	>95
East Midwest (IL, IN, West KY)							
85 FGD	2000 (350)	3.5	25 (22)	40 (35)	60	70 (64)	95 (95)
90 FGD			75 (70)	90 (93)	95	>95 (100)	>95 (100)
95 FGD			>95	>95	>95	>95	>95
National total							
85 FGD	8830	16.1	70	80	85	90	>95
90 FGD			90	95	>95	>95	>95
95 FGD			>95	>95	>95	>95	>95

*Equals 1.2 lb. SO_2 per million Btu monthly average.

Source: Based on Bureau of Mines data on U.S. coal reserves and the sulfur reduction potential of U.S. coal reserves. Parenthetical data provided by the National Coal Association. For North Appalachia, NCA data reflect Ohio and Northern West Virginia only.

full, partial and regional standards—and presented modeling results. Table 6-2 was presented to show regional emission impacts. At the briefing, Stuart Eizenstat represented the Domestic Council, Secretary James Schlesinger from DOE, Cecil Andrus from Interior, Alfred Kahn, the chief inflation fighter, and Charles Schultze from the Council of Economic Advisors attended, among others. EPA was working to secure Interior and OMB support for a strict standard to balance the cost-cutting point of view of DOE, DEA, and COWPS. The emission ceiling was not presented to this group, as an issue, just the options of full, partial, and regional scrubbing.

While some EPA technical staff favored a 0.55 lb. annual emission ceiling as most cost-effective, Walter Barber and Douglas Costle were uneasy about the impacts of such an emission ceiling on Midwest and northern Appalachian coal interests as shown by EPA microeconomic analysis. Nonetheless, they decided not to dismiss the 0.55 lb. annual emissions ceiling option publicly. The DOE, CEA and some EPA

Table 6.2: 1995 Regional SO_2 Emissions from Power Plants

East	
Continue current NSPS	11.2
Partial control	9.5
Regional standard	9.5
Full control	9.5
Midwest	
Continue current NSPS	8.1
Partial control	7.7
Regional standard	7.7
Full control	7.7
West South Central	
Continue current NSPS	2.6
Partial control	1.8
Regional standard	1.8
Full control	1.7
West	
Continue current NSPS	1.7
Partial control	1.2
Regional standard	0.9
Full control	0.9

Source: Costle Briefing Materials: Revised NSPS for Power Plants; SO_2 Control Alternatives, March 14, 1979.

economists who were most interested in a cost-effective standard (called the "decision optimizers") favored the very low ceiling as the cheapest way to achieve desired air quality, encouraging the burning of low-sulfur coal instead of installing costly technological controls on high-sulfur coal. But Costle had to be concerned also about political support for the final standard and did not want his Agency to approve a very stringent ceiling that would have major disruptions in coal markets. Further, Costle read the Clean Air Act as preventing major harm to high-sulfur coal producers.

NATIONAL COAL ASSOCIATION "BLITZKRIEG"

About this same time in March that EPA policy-makers were rejecting the 0.55 lb. emission ceiling among themselves, the National Coal Association in Washington became alert to EPA staff's serious consideration of a very low emission ceiling. EPA staff had sent its analyses to NCA previously and received no response. Now, NCA took notice, conducted a quick survey of its member coal companies, and became alarmed at what they concluded about a low emissions ceiling. Coal in the 6 to 9 lb. SO_2/mill. Btu range would be excluded with 90 percent reduction of SO_2 and an emission ceiling of 0.55 lb., they concluded.[93] Much of this coal is owned by Peabody and Consolidated Coal Companies, who are major actors in the National Coal Association (NCA).

On the issue of full vs. partial scrubbing, NCA had elected to remain neutral, since it represented both high- and low-sulfur coal companies, that were divided on the percentage reduction issue. The emission ceiling was another matter, since Western low-sulfur coal companies had no strong feelings about the ceiling, but high-sulfur interests felt they would be seriously hurt by a ceiling lower than the 1.2 lb./mill. Btu ceiling which EPA had proposed in September.

NCA concluded that EPA's analyses understated the negative impact of a lowered ceiling on high sulfur coal in three ways:[94]

1.*Coal reserves data*. The ICF model of impacts on coal reserves in key Midwestern and Eastern states used Bureau of Mines data on national coal reserves. These are total reserves and not those readily available for mining and use by utilities. BOM data included metallurgical low-sulfur coal which would not be purchased by utilities, that which is under fragmented ownership, coal under cities and highways, and located in difficult geological terrain such as steep slopes, faults, and in oil and gas wells. For example, NCA noted that 70 percent of the Illinois reserve base was unusable, according to the Illinois State

Geologic Survey. NCA said EPA should use, instead, the reserves currently owned or leased by coal companies, the reserves committed to development in the near term. These tended to be higher-sulfur varieties.

2. *Coal washability.* NCA felt that EPA's analysis overstated the degree of sulfur reduction achievable through coal washing. EPA had assumed that 5 percent of SO_2 could be washed from low-sulfur coal and 35 percent from high-sulfur coals. NCA's survey of its members in western Kentucky, West Virginia, Ohio, Illinois, and Indiana showed that actual operating practices at coal preparation plants removed from 20 to 31 percent of SO_2 from high-sulfur coals, averaging 27 percent. Thus, EPA overstated washability by about 8 percentage points.

3. *Utilities' attitudes and coal buying practices.* NCA described utilities' attitudes toward pollution control as conservative. Companies would design a scrubber to meet the NSPS and then buy lower-sulfur coal than required to meet the emission limits, so that they would have a margin of safety against pollution violations. Utilities had little faith in scrubbers' efficiency and would seek to protect themselves during times of malfunction or in the event sulfur content of coal should vary more than anticipated. This conservative attitude would even further reduce demand for high-sulfur coal beyond what EPA projected. The ICF model in use by EPA was a cost-minimizing model that assumed if it were cheaper that utilities would buy a better scrubber and then burn higher-sulfur coal. The model did not account for reluctance of the utilities to take risks, but assumed the lowest cost decision would be practiced.

Walter Barber called a meeting for April 5 for all interested parties to discuss the coal reserve data and the use of physical cleaning credits. Thirty-four persons attended, including representatives of NCA, various coal companies, Natural Resources Defense Council, Environmental Defense Fund, United Mine Workers, UARG, DOE, and EPA. EPA presented results of its microeconomic analysis of coal market impacts. NCA presented data on the sulfur content and washability of reserves that are currently held by NCA companies. While the reported NCA reserves represent a very small portion of the total reserve base, they indicate reserves which are planned to be developed in the near future. From this survey NCA differed with EPA's coal impact number, saying that the impacts on high-sulfur coals was much more serious than EPA estimated. (See Table 6-1 for a comparison of NCA numbers with EPA's conclusions on percentages of burnable coals under alternative ceilings.) However, Barber pointed out that NCA numbers were within five percentage points of EPA's in many instances, which EPA staff felt confirmed the results of its analyses, and made the Battelle model acceptable for studying coal reserve impacts.[95]

On April 6 Bagge wrote Costle with more detailed NCA survey data on sulfur and washing potential of coal, concluding:

The enclosed data documents two significant points:

- The devastating impact certain alternative sulfur emission ceilings could have on coal reserves.
- The serious concern we have about the validity of the across-the-board 35 percent sulfur removal by washing assumption relied on by your staff in their studies of the sulfur emission ceiling issue.

We believe the significant national policy consequences that will result from such an EPA proposal deserve serious and full examination by the public, industry, labor and the full range of government, Congressional and state interests."

NCA projected impacts of a lowered ceiling to be:

- There would be little or no new mining in some important coal producing states.
- Considerable time would be required to acquire new reserves, *if they are available.*
- The cost of acquiring reserves would be much higher than reserves now owned.
- Mining costs would be much higher because the reserves are typically thinner and at greater depth.

Miner Unemployment Situation. Large numbers of miners are now unemployed or working short work weeks because of the slack demand of coal. The potential lowering of emissions limitations would make this situation even worse in the East and Midwest.

- Coal demand has grown slowly, due in large part to government requirements which have made coal costly, difficult, or impossible to use or have made oil less costly;
- The number of miners in the East and Midwest now out of work and/or working short work weeks include:

  In West Virginia, 7,000 miners are unemployed and another 1,500 to 2,000 are working short work weeks

  In Ohio, 2,000 miners are unemployed and another 2,600 are working short work weeks

  In West Kentucky, Indiana and Illinois, 1,800 miners are unemployed;

- Electric utilities—and therefore, consumers of electricity—would be denied access to some of the most economically recoverable coal reserves if unnecessarily strict emissions limitations are set;
- The technological and regulatory risk associated with the use of scrubbers tends to drive utilities to use coal with the lowest possible sulfur content—rather than running the risk of having a plant become unavailable. This will put more upward pressure on the price of coals that would be usable under tighter emission limitations;
- Ideally a diversity of supplies of coal would be available to utilities—from surface or deep mines in various parts of the country. Weather conditions and tranportation limitations, for example, suggest the need for considerable diversity; and
- Consideration of utilities' dependence on fewer coal producing areas would likely lead to greater increases in fuel costs than is suggested in EPA's analyses. For example there would be:

    Greater competition for scarce coal resources
    Greater demand on available transportation capacity
    Greater exposure to State Severance Tax increases.[96]

EPA RESPONDS TO COAL INDUSTRY DATA

In general, EPA staff felt that the NCA data confirmed their basic data base. They did realize as a result of the information NCA presented that they had made two mistakes on fine points of the analysis. They redid the analyses with corrections, and low emissions ceilings were shown no longer to be favorable. Substantial impacts on coal mining were shown.

The first of the two changes made was to reassess the state-of-the-art of coal cleaning technology. Second, in order to fully explore the potential for dislocations in regional coal markets, EPA adopted a more conservative picture of utility coal buying practices. The Agency

> Concluded that actual buying practice of utilities rather than the technical usability of coals should be considered. That analysis identified coals that might not be used because of conservative utility attitudes toward scrubbing and the degree of risk that a utility would be willing to take in buying coal to meet the emission ceiling.[97]

The Battelle modeling microanalysis was performed, but with two new assumptions:

> (1) Utilities would purchase coal that would provide about 10% margin below the ceiling in order to minimize risk; [and] (2) Utilities would purchase coal that would meet the ceiling (with margin) with a 90 percent reduction in potential SO_2 emissions. This assumption reflects utility preference for buying washed coal for which only 85 percent scrubbing is needed to meet both the percent reduction and the ceiling requirement. In contrast, the previous assumption had been that utilities would do 90 percent scrubbing on washed coal—a practice which results in a greater than 90 percent reduction in potential SO_2 emissions. This analysis was performed using EPA data and (for comparison purposes) the NCA data at 1.0 and 1.2 lb. monthly ceilings.
>
> For the 1.0 ceiling, the results showed that 20 to 30 percent of the coal reserves in Ohio, Illinois, Indiana, Western Kentucky, and northern West Virginia would be bypassed by conservative utility buying practices. To assure that the NSPS does not seriously constrain Eastern high-sulfur coal production, the standard would have to require 88 percent scrubbing (92 percent reduction in potential SO_2 emissions) or provide a 1.2 lb. SO_2/mill. Btu emission ceiling.[98]

SHIFT AT EPA FROM FULL TO PARTIAL SCRUBBING STANDARD

While this flurry of activity was occurring over the impacts of ceilings on coal producers and miners, pressures were mounting on Costle for his decision on the final standard. EPA lawyers had gone to court to secure an extension of the March deadline for publishing the final standard. The court granted an extension to June 1 and directed Costle to give guidance to his staff by April 16 about his decision. The court would grant no further extensions and would not allow the public record to be reopened for additional comment.

Other pressures came from active press coverage of EPA's decision process. Since the September NSPS proposal, journalists from the daily press—and particularly trade magazines and environmental publications—reported changes from week to week in EPA staff thinking about the standard. Preferences by technical staff for 0.55 lb. ceiling or a 0.8 lb. ceiling, for a variable percentage reduction or uniform scrubbing, were written as "EPA's newest position."

Environmental groups' suspicions mounted about EPA's objectivity as stories were carried about coal and utility industries' involvement at EPA, in White House and Congressional meetings and about EPA's changing positions. Also, inquiries from Capitol Hill were increasing at EPA about the forthcoming decision.

Within EPA, Drayton, Gamse and their staff were pushing hard for partial scrubbing, based on these considerations:

In summary, choosing a full scrubbing standard:

- Significantly reduces Western power plant emissions, but smelter emissions dominate regional loadings in the West;
- Costs $1 to $1.7 billion more in annual utility spending than partial scrubbing;
- Would increase oil use about 200,000 barrels per day more than partial scrubbing;
- May preclude development of promising and more economical pollution control technologies.

A regional standard: (full scrubbing in the West, partial elsewhere)

- Reduces Western power plant emissions as much as full scrubbing;
- Costs about $300 to $600 million more in annual utility spending than partial scrubbing, but the cost per ton of SO_2 removed is the same as partial scrubbing;
- Eliminates the difference in oil use compared to partial scrubbing;
- Encourages development of dry scrubbing and other new SO_2 control technologies;
- [Is] difficult to write, legally, however.

The partial scrubbing standard:

- Leads to Western power plant emissions which are a third higher than under the regional or full control standards;
- Imposes lower SO_2 removal costs on new power plants than full scrubbing, but the cost per ton of SO_2 removal is still about two to four times the costs of other currently required SO_2 reductions;
- Forces us to rely more heavily on the states and the PSD program to control Western power plant emissions;
- Encourages development of dry scrubbing and other new SO_2 control technologies.[99]

Drayton, Gamse and OPE staff pointed out that while emissions in the West would decrease under either partial or full scrubbing as compared to the existing standard, the decrease would be 200,000 tons more with full scrubbing than partial. However, they were quick to point out, this difference was only about 10 percent of the total 1975 SO_2 emissions in the region. The big polluters were Western smelters which received pollution control exemptions in the Clean Air Act. If EPA wanted a significant SO_2 improvement those sources should be the focus, OPE contended.

Tundermann wrote:

> "What happens to Western emission in 1995 will be dominated by what happens to smelters." As Figure 2 shows, the difference in 1995 between full and partial scrubbing (0.3 million tons) is overshadowed by expected reductions in smelter emissions (1.2 to 1.7 million tons).[100]

By spring, Costle had reached a neutral, middle-ground position, not strongly favoring either a full or partial control standard. While he wanted a tight full scrubbing and felt CAA presumed full scrubbing, analyses were showing that this may not be the best standard. Costle was impressed by the $1 billion a year saved by partial scrubbing over full scrubbing, a $10 billion present value savings. Furthermore, model-

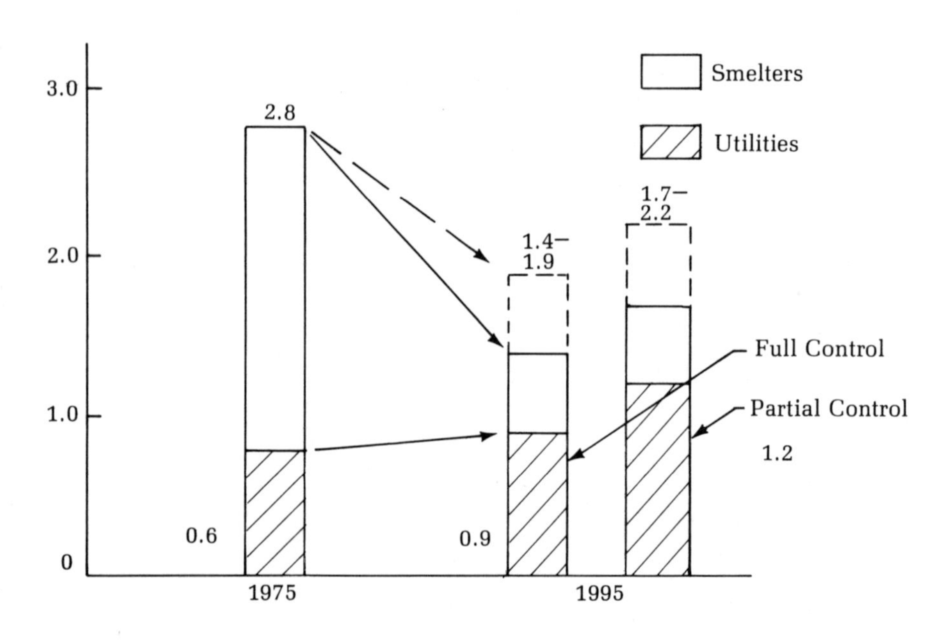

Figure 2. Comparison of Western SO_2 Emissions from Smelters and Utility Boilers.
(Assumes smelters will not continue to receive extensions, and will develop and install control techniques.)

Source: David W. Tundermann memorandum of March 9, 1979.

ing was showing no major reduction in national emissions from full control: 19.8 million tons emitted in 1995 with full control and 20.3 million tons for partial control. Indeed, the Midwest and East were going to experience less emissions with partial than full control.

In areas of the West where emissions were greater with partial than with full control the PSD rules could be used to require higher than NSPS limits to protect Western visibility, should that prove to be necessary. But even the visibility experts were not sure that any major impact would occur from the increased emissions which partial scrubbing would generate.[101]

Costle was coming to think of the partial control standard as a balance between environmental objectives and the economic and energy consequences of the standard. The administrator was performing the balancing act required by the Clean Air Act and necessitated by national environmental politics.

Also, Costle felt the need to save money from the September proposal, as a matter of symbolism, to show White House officials who were concerned about inflation and the public that EPA really was sensitive to costs of pollution control. Costle knew that with a price tag of $33 billion for utility air pollution abatement, it would strengthen his case to say he had cut $10 billion from the total.

By spring Barber and other OAQPS staff reversed their previous preferences for full scrubbing, based on modeling results which showed partial scrubbing more cost-effective than the September proposal.

DRY SCRUBBING TO THE RESCUE

With Costle seriously considering approving partial scrubbing—the key policy issue—there remained the more technical task of deciding percent reduction figures for SO_2.

Costle, Hawkins, and Barber did not want to accept DOE's position on percent reductions of 33 to 85 percent and certainly not industry's position of 20 to 85 percent. EPA air program staffers were considering percent reductions of about 50 to 90 percent, averaged monthly. The lowest figure for minimum scrubbing that EPA considered was 50 percent removal.

Then, a rapidly developing technology, dry scrubbing, provided the basis for a minimum percent reduction figure of 70 percent, and clinched Costle's support for a partial standard over full scrubbing.

In dry scrubbing a wet atomized slurry is introduced into the flue gases, dried by the hot flue gas, and collected as dry particulate. Dry scrubbers could be used effectively on coals with less than 3.0 lb. of

SO_2/mill. Btu.[102] If utilities would use dry scrubbers on cleaner low-sulfur coals to comply with a partial NSPS, cost differences between full and partial scrubbing would increase significantly. Dry scrubbing can reduce control costs because a utility could install one piece of equipment for both SO_2 and particulate control, instead of two. Also, there is no wet sludge waste product to dispose. If one assumed widespread use of dry scrubbers in 1995 with a partial scrubbing NSPS, the difference in annualized cost between partial and full scrubbing was estimated to increase from $1 billion to $1.7 billion. The difference in present value between full and partial scrubbing increased from $10 billion to $17 billion.[103]

However, dry scrubbing is not as efficient as wet scrubbing. Removal efficiencies were estimated by EPA's research staff to be about 50 to 70 percent, compared to 90 percent removal of SO_2 for wet scrubbers. Dry scrubbers can remove 50 to 70 percent of the SO_2 (in low-sulfur coals only) for the same or lower cost than a wet scrubber capable of 33 percent removal (the minimum requirement under some partial scrubbing options). Thus, the availabilities of dry scrubbers could permit a higher minimum removal requirement (50 to 70 percent) to be specified in a partial scrubbing standard than is practical if wet scrubbing is the assumed technology. This tighter requirement could yield more SO_2 removal at the same cost than with wet scrubbers.

This was not the first time that dry scrubbing had been discussed, but it had not previously received substantial attention. Back in January 1979 EPA's Office of Research and Development had suggested that the new pollution control technology could save the utilities money. Dry scrubbing had been noted as an emerging technology in the original September proposal. Also, public comments from industry supported a sliding scale standard, so that dry scrubbing would not be precluded. Now, dry scrubbing began to receive high priority attention.

David Shaver in an April 11 memo to Bill Drayton described a dry scrubbing standard for low sulfur coals:

> *An alternative method to write this standard and allow less than 90% control on low-sulfur coal is to define the "best available demonstrated technology" (BADT) differently for low- versus high-sulfur coal.*
>
> It is conceivable that EPA could write a "full scrubbing" standard which would allow less than 90 percent BADT as 50 to 70 percent removal on low-sulfur coal, based on dry scrubbing technology, and in defining BADT as 90 percent on high-sulfur coal, based on wet scrubbing technology. Considerations of cost and environmental impacts should allow this determination.

Problems with this approach arise due to the development stage of dry scrubbing and the lack of opportunity for public comment regarding the technology. We would need to make a determination that dry controls are "demonstrated," which ORD seems reluctant to do. Also, OGC believes that a standard based on dry controls would require reopening for *comment*, which the consent agreement forbids in its current form. (italics mine)[104]

PHASE 3 MODELING

The last phase of computer modeling occurred between the December public hearing and May, 1979 when the standard became final.

During the last phase of modeling, dry scrubbing on low-sulfur coal was introduced as an assumption in modeling, considerably lowering estimated control costs. This was the only variable that changed during this phase of modeling. The results strongly favored the partial control standard, based on dry scrubbing efficiencies (70 percent SO_2 reduction). Emissions were lower and costs were lower. Modeling showed that the partial scrubbing standard (using 70 percent minimal removal with dry scrubbing and 90 percent for high-sulfur coal with FGD) was more cost-effective than the full scrubbing option which EPA had proposed in September.

The only alternative that would have been more cost-effective than a 70 to 90 percent sliding scale standard with dry scrubbing was the option of a 0.55 lb. annual average emission ceiling and specified percent reduction or widely varying reduction. A range of emission ceilings was considered in modeling during this phase, along with various percentage reduction figures. Fourteen alternatives in all were modeled.

The regional differences in emissions projected by the model remained. In the East, partial control with 70 to 90 percent SO_2 reductions produced the least emissions, less than DOE's 33 to 90 percent variable standard and less than full scrubbing. Nationwide, emissions were reduced by 10 to 13 percent over the 1971 NSPS. The Midwest emissions are only slightly reduced. But in the West and West South Central regions, where significant growth is expected and today's emissions are relatively low with lower-sulfur coals or natural gas, full control would reduce current emissions by 40 percent; 33 to 90 percent partial options reduce emissions by 30 percent. Under the 70 to 90 percent partial standard, which EPA finally adopted, the West will experience 200,000 more SO_2 emissions than under full control as proposed by EPA in September.

COSTLE APPROVES DRY SCRUBBING

While dry scrubbing had been discussed among EPA technical staff, Hawkins had preferred not to suggest dry scrubbing or any other partial control solution to Costle as long as Hawkins thought there was a chance the administrator might still favor full scrubbing. But now Costle was strongly leaning toward partial scrubbing and Hawkins asked ORD to determine the maximum removal dry scrubbers could achieve. In April, at a meeting of the AAs and the administrator, Bill Drayton urged adoption of the partial scrubbing standard based on dry scrubbing for low-sulfur coals and FGD for high-sulfur coals as the most cost-effective and politically balanced standard.

When dry scrubbing was presented to Costle as a way to set minimum scrubbing efficiencies, he approved it enthusiastically. To Costle, dry scrubbing had both the appearance and reality of balancing environmental, economic, and energy objectives. A 70 to 90 percent reduction standard with dry scrubbing would reduce emissions more than the DOE and industry options and more than full scrubbing and at much less cost. Dry scrubbing provided a technically valid means to raise minimum scrubbing above DOE and industry proposals. It was an innovative control technology that would be encouraged by EPA's sanctioning its use in the NSPS. Dry scrubbers are a less complex technology that provide potentially greater reliability than wet scrubbers and require less energy to operate. Dry scrubbing would save scarce and precious water in the West where dry scrubbers would be used mostly on local low-sulfur coals.

Furthermore, no wet sludge needs to be disposed of and the dry wastes are easier to handle. There was something in this choice for environmentalists, the energy conscious, and inflation fighters.

Costle and Hawkins both approved the 70 to 90 percent partial scrubbing standard,* based on dry scrubbing.

After making his decision to support a 70 to 90 percent reduction standard based on dry scrubbing for low-sulfur coals and FGD for high-sulfur coals, Costle kept his decision "close to the vest."[105]

The environmental newsletter *Air/Water Pollution Report* was left to speculate on April 23

---

*For public relations purposes, EPA began to call this partial scrubbing standard a "variable standard," which was the language that DOE and the utilities had used. "Partial" scrubbing, EPA reasoned, had the tone of backing away from the tougher sounding "full scrubbing" standard proposed in September.

Costle was described as "agonizing" over the NSPS issue last week after a four-hour April 13 meeting with key deputies and subsequent internal discussions the following weekend. But some EPA and DOE officials were convinced last week that Costle had indeed made up his mind—even if he was keeping program officials in the dark, as some suspected, to stem press leaks.

Sources believed it inevitable—unless something changes Costle's mind at the last minute—that he would ultimately opt for some form of partial scrubbing, favored by DOE, White House inflation fighters, and utility interests but vehemently opposed by environmentalists . . . DOE had called for a minimum 33 percent SO_2 removal standard for very low-sulfur coals, but EPA sources indicated the environmental agency might select a 50 percent or even higher minimum standard to curb excessive emissions in the West.[106]

LAST MANEUVERS: SENATOR BYRD AND THE EMISSIONS CEILING

As EPA was coming down to the wire in setting its final NSPS, with Costle having decided percent reduction issues but not the emission ceiling, Senator Byrd became actively concerned about the possible impacts of a lowered ceiling on coal production in his state of West Virginia. Unemployed coal miners in his state needed no additional troubles from EPA, he concluded. The National Coal Association had solicited Byrd's involvement to reverse what it considered to be EPA's preference for a 0.55 lb. emission ceiling. In fact, that preference was among technical staff, but carried in press accounts as EPA's official position. Barber was considering a ceiling about 0.8 or 1.0 but a 1.2 lb. emission ceiling (monthly average) was also under active consideration in modeling runs. Costle was leaning toward a 1.0 or 1.2 lb. monthly ceiling, because a lower limit would create too great a negative impact on high-sulfur coal markets.

On April 23 Senator Robert Byrd of West Virginia called the first of two meetings in his congressional office with EPA about the emissions ceiling. The meeting was attended by Byrd, NCA representatives, Stuart Eizenstat, Costle, and his staff. David Hawkins reported for the NSPS docket that NCA presented material it had given EPA opposing a ceiling more stringent than 1.2 lb./mill. Btu monthly average ceiling, as precluding substantial reserves in the East and Midwest. Hawkins reported for the written record that

Costle informed the participants at the meeting that it was not the agency's intent to rule out large portions of Eastern or Midwestern coal reserves as a result of setting the power plant new source performance standard. Stu Eizenstat confirmed that this was not the administration's intent either. . . . Senator Byrd expressed his opinion that a decision to set an emission ceiling which would preclude the use of coal reserves would cause him great concern.[107]

EPA STRATEGY FOR WHITE HOUSE REVIEW

While not yet deciding the ceiling issue, Costle, Hawkins, and their chief aides devised a strategy for final White House review of its preferred variable NSPS. EPA's recent acrimonious dealings with White House staff over promulgation of a revised national ambient air quality standard for ozone had persuaded Costle of the need to control the White House review process more effectively with this NSPS. Administration inflation fighters, backed by Eizenstat, had been concerned about the costs of the ozone standard and tried to interpret health impacts data to show that a more lenient standard was justified. Costle had told Eizenstat that he would have to go to the president to reverse EPA's ozone decision. The appeal was not made, but hard feelings remained after promulgation. Eizenstat felt that EPA had not been entirely forthcoming about the impacts of the ozone standard, and EPA officials felt that the White House staff was addressing technical issues, particularly health effects of alternative standards, for which they lacked technical expertise.

Costle wanted White House backing of his NSPS and did not want to repeat an acrimonious process. Another consideration was the recent White House involvement on the OSHA cotton dust standard, in which the president had made the final regulatory decision and received considerable criticism. Costle figured that the president might not want to make another such controversial and expensive regulatory decision, but that EPA should advise him of their forthcoming decision and give him the opportunity to decide the NSPS standard if he so chose.

The strategy was to hold an information briefing for the president, not a decision meeting. Key administration officials would attend, and speak. The president would have the opportunity to issue directives if he chose, or he could merely be briefed and allow EPA to finally decide the matter, keeping the decision away from his desk.

The meeting was set for April 30. To the news media on the outside, the White House meeting was expected to be a showdown. One account speculated this way:

Going to the mat on NSPS? EPA Administrator Douglas Costle is scheduled to attend what may be a showdown White House meeting tomorrow on EPA's controversial new source performance standards for coal-fired power plants. By late last week, EPA program officials were still uncertain precisely what NSPS recommendations Costle would take into the parley. Yet there was considerable speculation, both within and without EPA, that Costle would emerge endorsing a "partial" scrubbing plan, though likely a more restrictive version than those favored by the Energy Department and the utility industry. Sources predicted that Costle would try to offset this standard's relaxation by backing a tighter sulfur dioxide emissions "ceiling" than the one he proposed in September, but they also predicted that the ceiling would be less stringent than that favored by some segments of EPA's Office of Air Quality Planning and Standards.

These (EPA) officials predicted Costle initially would also back a monthly SO_2 ceiling equal to an annual average of 0.55 lb./mill. Btu, but might agree to a less restrictive monthly standard of 0.8 lb./mill. Btu (equal to a 0.7 lb./mill. Btu annual average). EPA's OAQPS had until recently supported—and some of its staffers still favor—the tighter standard, since it was projected to permit fewer emissions and be more cost-effective. But sources said OAQPS chief Walter Barber was persuaded at a recent meeting with coal industry officials that the stricter ceiling might preclude burning of significant quantities of high-sulfur Midwestern coal. Even EPA officials who disagreed with this assessment said Costle would not want to "convey the image" of being anti-coal, and added that a 0.8 lb./mill. Btu monthly level would still be somewhat tighter than EPA's September proposal.

Unmoved by coal industry arguments were environmentalists, who were fuming late last week at the prospect that Costle would adopt not only the utility industry's partial scrubbing and 30-day averaging time concepts, but would also heed the coal companies' call for a laxer ceiling. In a last-ditch effort to avert defeat on all these issues, the Environmental Defense Fund and the Natural Resources Defense Council wrote to Costle last Friday, imploring him to consider their arguments that full scrubbing and a tighter ceiling would be preferable from an environmental standpoint. EDF's Robert Rauch and NRDC's Richard Ayres contended that high-sulfur coal could still be burned under a tougher ceiling, since, they said, advance scrubbing technologies and coal washing could remove more than the 90 percent maximum SO_2 reduction that Costle was said to favor. (parens mine)[108]

PRESIDENTIAL BRIEFING

Costle briefed the president and top administration officials, including Eizenstat, Schlesinger, Kahn, and Schultze, presenting alternative

NSPSs with expected impacts of each. He indicated his preference for a partial scrubbing to 70 to 90 percent SO_2 reduction. About emission ceilings, he said he was considering the 1.0 and 1.2 lb. emission ceilings but had not resolved the issue, awaiting further modeling results. Attendees expressed concern that major dislocations in coal markets not occur for economic and political reasons. White House economists expressed support for a low ceiling as the most cost-effective approach, as shown by modeling.

Schlesinger presented DOE's formal position, favoring a 33 to 85 percent sliding scale, but the Secretary did not strongly oppose EPA's preferred 70 to 90 percent standard. Indeed, Schlesinger privately thought that Costle had reached a reasonable and balanced solution and was delighted with the conclusion.

Schlesinger attended the briefing with his assistant secretary Alvin Alm, not John O'Leary, a fact that was read by EPA officials as they entered the meeting as a sign that DOE would not vehemently oppose EPA's standard. Alm was considered more supportive of EPA than his DOE colleague O'Leary, who had been a strong EPA opponent on the NSPS issue. Indeed, at the meeting Schlesinger did not argue about the basic impact numbers, such as expected costs or emissions, nor about the quality and nature of the modeling analyses. Rather policy issues were discussed.

Charles Schultze stated his views, favoring a sliding scale standard of 50 to 85 percent SO_2 reduction, but did not aggressively call for a weaker standard than EPA had favored. No one was "going to the mat" with Costle over the NSPS.

The president, obviously well prepared for the meeting, asked Costle questions about dry scrubbing. He did not, however, give EPA any directions about the standard and as EPA officials left the meeting, they felt they had received tacit approval to reach their own conclusions on the NSPS.

Schlesinger and others had doubts also about dry scrubbing, which appeared to be a technological quick-fix, solving EPA's desire to raise the minimum emission floor above that suggested by industry and DOE. Such quick fixes are rare in policy decision-making and EPA had produced this one in a few weeks. To DOE and White House officials it seemed too good to be true.

Following the presidential briefing, Eizenstat checked into EPA's contentions about dry scrubbing. Within the next few days EPA's ORD staff briefed the White House staff about the control technology, and White House staff called pollution control experts around the country to verify dry scrubbing capabilities. EPA's position on dry scrubbing was confirmed.

SECOND BYRD MEETING

Senator Byrd called his second meeting about the ceiling for May 2. At the gathering attended by Byrd, DOE's O'Leary, Cecil Andrus, and NCA and EPA officials (Interior Secretary Andrus attended to discuss his Bureau of Mines reserves data), Costle described the emission ceilings EPA was considering and the impacts as shown by ICF modeling and EPA's own microanalysis of the high-sulfur coal regions. Hawkins' report to the NSPS docket was brief:

> The meeting was for the purpose of sharing our assessment of the Bureau of Mines' information and National Coal Association's information on coal reserves with Senator Byrd and to describe EPA's own emission ceilings which were under consideration. The attached materials were supplied to the attendees of the meeting.[109]

Following the May 2 meeting, Byrd wrote President Carter of his grave concern over any ceiling more stringent than 1.2 lb. Some 25 senators from high-sulfur coal states signed the letter, as well. Leaving no possibility that his concern would not be registered with the president and demonstrated to his constituents, Byrd then met with the president on May 8 to state his opposition to a lowered emissions ceiling. Several other senators accompanied him.

Still, the White House sent EPA no directive. It continued to be Costle's decision to set the ceiling and other aspects of the NSPS. Alternative ceilings were being fed in rapid fire fashion to the model to predict impacts.

The *Washington Post* was tracking events and on May 5, its lead story headlined, "EPA Will Relax Pollution Rules for Coal Power." "Relax" referred to the proposed partial scrubbing standard Costle favored as compared with the September full scrubbing proposal and to the 1.2 lb. ceiling Costle was considering as compared to the 0.55 lb. emission ceiling some staff had considered. The story reported events this way:

> The Environmental Protection Agency, under extreme pressure from the coal industry and Senate Majority Leader Robert C. Byrd (D-W. Va.), has decided to relax proposed air pollution standards for new coal-fired power plants.
>
> The decision, which has not been publicly released, represents a complete "cave-in" to industry and a betrayal of President Carter's commitment to clean air, environmentalists charged yesterday.
>
> But a coal company spokesman said industry generally is pleased

with the decision because it represents a "balance" between Eastern and Western coal interests and does not require excessive expenditures for cleanup equipment . . .

Administration sources said yesterday that EPA Administrator Douglas Costle has decided to require partial "scrubbing" of sulfur emissions on a sliding scale of 70 to 90 percent, depending on the sulfur content of the coal burned. His original proposal was to require all utilities to remove 90 percent of the sulfur.

Costle also has decided to retain the current emissions ceiling of 1.2 lbs./mill. Btus, measured as a monthly average. EPA had been considering a standard almost twice as strict.

The ceiling decision came after two weeks of what one Senate source called "hard-ball arm-twisting" by Byrd and other coal state senators. Byrd summoned Costle and White House adviser Stuart Eizenstat, strongly hinting that the administration needs his support on the strategic arms limitation treaty (SALT) and the windfall profits tax, according to Senate and administration sources. . . .

Environmental Defense Fund attorney Robert Rauch, calling Byrd's role "outrageous," said he will file suit. "A powerful senator is given information by a private party and forces Costle to back down. It makes a complete sham of rule-making procedures, which are supposed to be aboveboard."[110]

Reading such reports* and being highly suspicious about regulatory agencies meeting privately with spokespersons for regulated industries, the environmental groups would conclude that Costle was "caving in" to congressional and industry pressure.

EPA analytical results came in, showing only a slight increase in emissions in going from a 1.0 to 1.2 lb. ceiling, the two numbers Costle was considering. The 1.2 lb. ceiling would result in only an additional 50,000 tons of SO_2 nationally, over a base of 18 million tons in 1995.[111] Costle felt that even this small increase would not likely occur, since utilities would prefer lower-sulfur coal than the minimum permitted, because such coal is cleaner to handle and provides a margin of safety against pollution violations. A ceiling more stringent than the 1.2 lb.limit was shown to disrupt coal markets. Up to 22 percent of high-sulfur coal reserves in the East, Midwest, and portions of Northern Appalachia could be precluded from use if the emission limit were 1.0 lb./mill. Btu.[112] This coal market disruption would be counter to the Clean Air Act's call to encourage use of local coals.

*See Appendix 6 for additional media coverage.

EPA policy officials had to weigh this relatively small pollution addition from a higher ceiling against the substantial political problem that would result from taking on Byrd and the other high-sulfur coal state senators. Dislocations in coal miners' jobs were another consideration. Nor would appearing to be anti-coal in times of energy shortages make the Carter administration look responsible. The weight of the decision was clearly with the higher ceiling. Hawkins recommended that Costle approve the 1.2 lb. ceiling, and the administrator agreed.[113] The views of Byrd and other members of Congress "at most, served to reinforce the administrator's own judgment that the proper level of the standard was 1.2 lb./million Btu,"[114] EPA concluded.

Costle explained his ceiling decision in two ways:

> I have decided to hold that limit at its current ceiling, that is, 1.2 pounds of sulfur dioxide per million Btu (British thermal units), measured on a 30-day average. I have done so for two basic reasons:
>
> - Our analysis of existing coal reserves has shown that significant portions of those reserves located in the Midwest and Appalachia would not be able to be burned under this strict new standard, unless the utilities chose to control emissions by more than the required 90 percent.
> - Just as important, keeping the 1.2 million Btu ceiling will *not* result in substantially greater regional emissions than the 1.0 pound ceiling I had been considering. The increase will be less than 50,000 tons of SO_2 in the East and Midwest over a base of 18 million tons.[115]

As to pressures from Senator Byrd and others, Costle dismissed the uproar, referring to criticisms about slight emissions increase as "a tempest in a teapot." He said, "It's gotten to be a symbolic issue. . . . On balance we've come out with a very tight standard."[116]

7

FINAL RULE

This chapter summarizes the final rule, EPA's assessments of its impacts, and legal reaction to the NSPS.

The final standard that Costle announced on May 25 and published in the *Federal Register* June 11, 1979,[117] called for:

SO_2

- Percent reductions varying from 70 percent on low-sulfur coals (assuming dry scrubbing) to 90 percent on high-sulfur coals (assuming FGD);
- Floor set at 0.6 lb./mill. Btu set at a level to accommodate dry scrubbing on low-sulfur coals. When emissions below 0.6 lb. SO_2/mill. Btu (30-day rolling average) a 70 percent reduction applies;
- Ceiling of 1.2 lb./mill. Btu;
- Average period: 30-day rolling average;
- Anthracite coal burning exempt from the percent reduction, but not the emissions ceiling.

NO_X

- Varies with fuel; same as September proposal;

Particulate

- .03 lb./mill. Btu; same as September proposal;

Monitoring

- Continuous monitoring, minimum data required is 22 days out of 30, 18 hours out of 24.

IMPACTS OF NSPS

Costle summarized the NSPS impacts:

First, protecting air quality: We anticipate significant reductions in emissions from new coal-fired power plants between 1983 and 1995. In terms of national totals, sulfur oxides emissions from new plants will be half of what the old standard allowed. Particulate emissions will be 70 percent lower, and nitrogen oxides will be 20 percent lower.

Second, achieving maximum reductions in a cost-effective way: By setting a variable standard based on the characteristics (i.e., sulfur content) of the fuel, we have permitted slightly less control of low-sulfur coal to allow use of a technology which promises to sharply reduce the costs of control. We calculate that the difference in sulfur oxide reduction between a requirement for full scrubbing (90 percent removal) and the variable standard I am announcing today will, in 1995, amount to less than 20,000 tons per year in the West.

In the East emissions are actually reduced more (400,000 tons per year) under the variable standard than full control because dirtier, older plants are replaced more rapidly.

Allowing the option of dry scrubbing on low-sulfur coals could save as much as $13 billion in the present value of 1995 annual utility expenditures.

Our projections indicate that the average monthly residential bill will increase by 2 percent in 1995 as a result of these standards. In addition, the average family will also incur an additional $1.20 per month increase as the result of indirect costs (increased prices paid for products they buy which reflect higher utility bills paid by manufacturers).

The increase caused by meeting the standard in the annual revenue needs of the utility industry will be only 2.5 percent of total benefits that the public will experience, such as cleaner air, better health, more room for necessary or desirable growth; and the preservation of visibility in the West.

Third, speeding up a promising new technology: Dry scrubbing systems are less complex than wet systems and therefore offer the prospect of greater reliability at substantially lower cost where lower-sulfur coal is used. Dry systems use less water, a real advantage in the West. They require less energy to operate. And the waste product, dry ash, is easier to dispose of in an environmentally sound way than is wet sludge.

Four, allowing use of our coal resources: There is a strong presumption in the 1977 Clean Air Act Amendments to encourage the use of regional and local coal. In the East, in particular, many of these coal resources are high in sulfur content. Because substitution of lower-sulfur coal was allowed before the 1977 Amendments, this meant that many Eastern utilities ordered coal from the West, where sulfur content was often lower. This resulted, among other things, in potential or actual job losses in eastern mines. Requiring full (90 percent) scrubbing of higher-sulfur coal while permitting 70 percent scrubbing of lower-sulfur coal will allow the development of our known resources of higher sulfur coal. By 1995, we expect that 30 percent less Western coal will be shipped East than would have been the case had the standards not been tightened.

Five, preserving our options for future growth: It is very difficult today to predict what our growth needs and desires will be 15, 20, or 30 years from now. We know, for instance, that recent years have seen a shift in population—with its attendant needs for new energy facilities—from the Northeast to the Sunbelt. This is one of the reasons that so much new electrical generating capacity is projected to be built in the West.

Faced with uncertainty about the nature or extent of that future growth, the prudent course is to keep our options open. We cannot afford to use up our capacity for sustaining any such growth in a sound way. Increasingly, we as a society are recognizing that clean air is a resource, a resource that needs to be allocated sensibly. The better the job that we do today in cleaning up our air and preserving a health margin for growth, the better off we and our successors will be several decades from now.[118]

(See Appendix 5 for impacts projected by computer analyses from Phase 3 of ICF modeling on which the above is based and which appeared when the final NSPS was published in the *Federal Register*.)

EPA "EXPECTS EVERYONE TO SUE" AND THEY DO

While EPA impacts analysis showed that the final NSPS would reduce emissions more than the full scrubbing standard in the September proposal, and at substantially less cost, the national news media generally described EPA's final rule as "weaker," "less stringent," and "more lax" than first proposed. Editorial comment was favorable, generally, calling EPA's decision a balance between coal, economics, and environmental considerations.

"The reality of the standard was considerably better than its image," Hawkins concluded.[119] Hawkins was quoted in print saying that he "expects everyone to sue"[120] EPA over the NSPS. Most did.

Environmentalists reacted with deep dismay to the final NSPS and readied their legal action against EPA. NCA's Carl Bagge said "We believe these standards to represent an honest attempt to balance future national environmental and energy objectives,"[121] which environmentalists took to mean he got what he wanted from the standards setting process. Senator Byrd also reacted favorably, which contributed to the impression that Costle had "caved in" to coal interests.

The last few weeks' flurry of meetings and modeling runs did not allow the time for review and interest group contributions that EPA had conducted earlier, which led environmental groups to believe they were being shut out consciously. Robert Rauch, attorney for the Environmental Defense Fund charged that the standard was developed under "raw political pressure" exerted by Senator Byrd.[122] Rauch agreed that the April 5 exchange of information on the coal reserves and several letters on the NSPS that were sent to Costle from UARG, the Coal Association, EDF and NRDC after the close of public comment period, were done in an open manner. Then, "suddenly everything went underground when Byrd and other senators became involved,"[123] Rauch said. "I think Byrd was sold a bill of goods," Rauch continued, "because the coal reserve information and the information given to him on scrubbing capability was wrong or exaggerated."[124]

EDF petitioned EPA to reconsider the standard, contending that "ex parte" communications between the Agency and Byrd and coal industry officials had swayed EPA's decision, and that the decision was made on the basis of new information raised after the close of public comment period in late January. He asked for reconsideration directed by Deputy Administrator Barbara Blum, who had not been party to the Byrd meetings.

The Sierra Club and the California Air Resources Board petitioned EPA for reconsideration of the whole standard, urging that uniform scrubbing be adopted and a much lower ceiling.

George Freeman, speaking for the UARG, was quoted as saying that the standard was one the coal states and utilities, "could live with."[125] However, UARG also petitioned EPA contending that a number of its provisions were too restrictive and impossible for the utility industry to meet, particularly the SO_2 standard.[126]

NCA asked the U.S. Court of Appeals for the District of Columbia for permission to intervene in EPA's side in the Sierra Club suit against partial scrubbing and the ceiling.

The U.S. Court of Appeals for the District of Columbia refused to hear the case until the groups had pursued appeals to EPA Administrator Costle. On February 6, 1980, Costle refused petitions from the EDF, Sierra Club, Kansas City Power and Light Company, Sierra Pacific Power Company, Idaho Power Company, the State of California

Air Resources Board, and UARG to reconsider the standard.[127] EPA responded that information used in the final decision was supplied for the docket and included in the record during the course of NSPS process. Impacts of emission limits on coal reserves "was not an issue because the agency wrote memos of significant communications and placed them in the public record. The Administrator's decision on the emission ceiling was not based on any information not in the public docket," EPA concluded.[128] Byrd had not exerted any undue influence on the decision. The technology and the numbers in the standard were valid and defensible, EPA concluded.

The principal modeling concept challenged by environmentalists was the "counterintuitive" conclusion that stricter emission controls will lead to more emissions and that less technological control meant more emission reduction. The debate turned on whether or not an increase in costs of constructing and operating a new plant would affect the utility operator's decision to retire an oil boiler, the dispatch of plants to make electricity, and the rate of construction of new plants. Environmentalists contended that the cost-minimizing feature of the model did not represent the true world and that political, financial, and institutional constraints would alter cost-minimizing behavior. EPA repeated its belief that cost-minimizing behavior is the most sound method of analyzing impacts of NSPS.[129]

The Sierra Club charged that the assumed oil price in the model was too low. EPA agreed that the current oil price was higher after the standard was final, although at the time modeling was performed it was correct. EPA performed sensitivity analyses and concluded that sensitivity runs "did not show significant changes in the relative impacts of alternatives, although total figures did change."[130]

Parties on all sides than asked the U.S. Court of Appeals to review EPA's decision and take up their appeals seeking revision of the NSPS.[131]

EPA and those challenging the NSPS have presented their arguments to the court and as of spring, 1981, no decision had been handed down.

In briefs to the court, the Sierra Club and the California Air Resources Board challenged the regulations on the merits, arguing that use of low-sulfur coal is not an appropriate method of control under Section 111 of the Clean Air Act. Rather, the parties claimed that Congress intended that all plants use the maximum practicable technological emissions control—wet scrubbers. The challenge argued that the administrator should have set a uniform percentage reduction requirement for sulfur dioxide rather than a variable requirement. EPA responded that the 70 to 90 percent reduction in SO_2 emissions reflected the best demonstrated technology for controlling such air pollutants, taking into account cost and environmental and energy impacts.

The Environmental Defense Fund challenged the regulations on procedural grounds, arguing that EPA's ex parte contacts with coal industry advocates after the close of formal rulemaking on January 15, 1979, "fatally flawed the rule-making."[132] The environmental group contended that EPA's six meetings with the National Coal Association, Senator Byrd, and other coal industry representatives were illegal, as were the letters from Senator Byrd and 25 other Senators and the last NCA and EPA studies of the impact of alternative rules on coal markets. Furthermore, EDF contended that DOE's O'Leary and other federal officials' communications were illegal, since these officials acted only as "conduits for coal industry ex parte comment to the EPA."[133] EDF contended that EPA was ready to adopt a stringent 0.55 lb. emission ceiling before the ex parte contacts occurred, based on EPA staff memoranda.

EDF contended that the 1.2 lb. emission ceiling was chosen by EPA as the result of political secret pressuring and had no basis in the public record. EPA's own staff analysis and ICF computer modeling showed the 0.55 lb. ceiling to be the most cost-effective, EDF said, and EPA staff was considering only alternatives with a 0.55 lb. ceiling.

EPA responded that EDF omitted from its account the EPA microanalysis of coal market impacts, which is a key part of the public record of the decision. The 1.2 lb. ceiling was based on the record as interpreted in light of congressional intent to avoid foreclosing use of Eastern high-sulfur coal. EPA's microanalysis showed a 1.0 lb. emission ceiling would preclude the use of up to 22 percent of the coal reserves in three eastern states but a 1.2 lb. ceiling would preclude less than 5 percent. Further, the environmental difference between a 1.0 lb. and 1.2 lb. standard was insignificant.

As to ex parte contacts, EPA responded that EDF was trying to impose upon the informal rulemaking procedures prescribed in the Clean Air Act, more formal procedures appropriate to adjudication. The comments and meetings with Senator Byrd and other members of Congress are highly appropriate, as are the meetings of EPA with other federal officials, the Agency claimed. The administrator made sworn statements that he always intended to select a ceiling which, like the 1.2 ceiling, would avoid precluding Eastern coals, because Congress so directed, and that he did not base his decision on Byrd's pressuring. "There was no impropriety in EPA receiving hundreds of letters from citizens, public interest groups, private companies and public officials. . . . Meetings within the Executive Branch before issuance of a final rule are entirely lawful and appropriate. . . . Communication within the Executive Branch is not only good law but also good government."[135] EPA could not agree with EDF that advice from a White House adviser or official is "inherently suspect or corrupt."[136]

The electric utilities appealed EPA's decision, saying that the 90 percent reduction of potential SO_2 emissions was not achievable. EPA countered that it was achievable, based on 92 percent median of SO_2 removal by flue gas desulfurization together with sulfur removal by coal washing and ash retention, taking into account the variability in FGD performance.

The utility industry also contended that the NSPS particulate standard was not achievable, that the standard was based on test data developed under conditions different from those specified for compliance. Therefore, the industry appealed, the technology is not demonstrated to be achievable at lignite-fired plants. The industry claimed that the Agency failed to adequately consider the cost of achieving the particulate standard. EPA responded that both baghouses and electrostatic precipitators could meet the particulate emission limit, at reasonable cost.

8

CONCLUSIONS: BALANCING ON A TIGHTROPE

This chapter draws conclusions about the standard, the rule-writing process and the key decision-makers. Lessons from this case for future environmental rule-making are offered.

EPA performed a balancing act in choosing among regulatory alternatives—weighing and trading economic, energy, and environmental objectives and the political interests they represent.

The result—the final NSPS promulgated by the federal agency—is a balance both in appearance and in fact. Both are essential while walking the tightrope of national environmental politics.

THE STANDARD IS BALANCED

In these ways the final NSPS is a balance of national objectives:

- On the issue of percent reduction of SO_2, the 70 to 90 percent reduction, averaged monthly, based on dry scrubbed low-sulfur coals and wet scrubbed high-sulfur varieties is a middle ground position. It lies between the utility industry position of 20 to 85 percent scrubbing and the 33 to 85 percent, averaged monthly, favored by DOE, on the one extreme, and that of the environmentalists and California Air Resources Board who favored uniform 95 percent removal, averaged daily, at the other extreme.
- On the issue of averaging times, the 30-day rolling average is a compromise accommodating enforcement needs with the engineering realities of control technology performance and the variability in the sulfur content of coal. Annual, 24-hour and 3-hour averages were the extreme positions. The continuous monitoring provisions are some of the most important features of the NSPS.

The ceiling of 1.2 lb. SO_2/million Btu, averaged monthly, is a consequence of economic and unemployment realities, as translated through the political system and is not likely to make much environmental impact, as compared to the 1.0 lb. monthly ceiling actively considered as an alternative. There is, however, considerable environmental difference in the impacts of the 1.2 lb. monthly ceiling chosen and the lower 0.55 annual (0.63 monthly) ceiling considered by technical staff. But the 0.55 annual average ceiling was always unacceptable legally and politically in Administrator Costle's mind. Environmental groups in appealing EPA's decision did not consider the difference in decision factors that prevail at the Administrator's level, as contrasted with mid-level staff and modelers.

As a practical matter, the significance of the 1.2 lb. ceiling rather than the 1.0 lb. ceiling alternative is in terms of EPA's image with environmental groups and the legal action it stimulated. Because of the rapid-fire actions that occurred among EPA officials, congressional and White House leaders and the appearance given through the media that EPA was rapidly changing its position on the ceiling, the impression was created that EPA cut back rooms with regulated interests. Consequently, the reality of the standard was better than its image, at least with the environmental community. While the public generally perceives the NSPS as balanced, it is seen as one tilted slightly in favor of utility and coal interests, as compared with the draft standard, a fact that was refuted by the macroeconomic analyses. Environmental groups, however, never accepted the modeling analysis. Furthermore, some environmental groups see no need to balance various interests in concluding a standard. Indeed, they see the Clean Air Act as mandating that EPA make purely a technical decision about the very best performance of demonstrated pollution control technology. The Act, then, requires EPA to mandate that conclusion for all sources, under all conditions, according to some environmental interpretations.

The balance as required by the Clean Air Act was struck by Administrator Costle based on these impacts:

- Emission reductions. Impacts on visibility in West, on populated Eastern cities, acid rain;
- Other environmental impacts, such as water use in West and solid waste disposal;
- Costs to utilities, costs to electricity consumers;
- Impacts on high-sulfur and low-sulfur coal markets and political consequences;
- Oil consumption and reliability of electrical energy;

- Limits and mandates of Clean Air Act (did the law require full scrubbing; how much should high sulfur coal markets be protected);
- Technological feasibility of scrubbers, baghouses;
- Enforceability. (How much continuous monitoring data were required statistically to prove compliance or noncompliance, for example, and how to achieve daily compliance data);
- Variability in sulfur content of coal;
- Attitudes of utilities toward scrubbers, unwillingness to take risks and the resultant impacts this would have on the high-sulfur coal markets (parens added)

In addition to its technical component, balancing had a political element. Costle and his agency had to appear and in reality be sensitive to various administration concerns including fighting inflation, developing coal use, saving oil, as well as environmental protection. The administration spokespersons for these issues had to have their views considered. Various interest groups' views also had to be considered and incorporated, as appropriate, in the balancing process. Costle could not be, nor appear to be, a one-issue administrator, protecting air quality at all costs.

The balance struck by Costle reflects his interpretation of the data and circumstances as described to him by his staff and many other sources in and outside government, whose views he solicited. Another EPA administrator might have struck a different balance, since several standards were possible. Another administrator might have tilted the balance inherent in the NSPS more with the regulated parties and approved a much more lenient minimum scrubbing requirement, or have sided with the environmentalists and mandated full scrubbing on all coals, with a much tighter ceiling. The decision reflects Costle's view of EPA and the direction in which his agency should be headed. Costle tried to steer EPA on a middle course between various interest groups and national priorities in the pursuit of environmental improvements. It was a delicate political course since EPA depends on the environmental community and its friends on Capitol Hill as a vital political base. Costle sought an environmentally sensitive course, but one that will allow the Agency's laws, budgets, and regulations to survive and thrive in the national political arena in times of strong competing claims of regulatory reform, energy needs, budget balancing, and inflation fighting. Balancing the competing environmental, energy and economic claims facing EPA at both technical and political levels is the daily business of the administrator.

THE STANDARD IS MORE COST-EFFECTIVE THAN FULL SCRUBBING

The final NSPS is more cost-effective than the full scrubbing standard proposed in September, given the assumption about the cost of scrubbers. This is due to older dirtier plants being left on line longer with full scrubbing. This fact was not apparent when a single model plant was studied but became clear when industry-wide impacts were assessed by way of the computer model. The final standard reduces emissions more and at 25 percent less cost than the September proposal. Both the 70 to 90 percent partial scrubbing option EPA chose and the 20 to 85 percent partial scrubbing favored by industry were shown by modeling to be more cost-effective than full scrubbing. The Clean Air Act did not permit the most cost-effective standard which was a low ceiling such as 0.55 lb. on an annual average with no percentage reduction, favoring the burning of low-sulfur coals.

The Clean Air Act was used by high-sulfur coal interests as a protection policy for high-sulfur coal markets. While the unusual political alliance of environmental groups and high-sulfur coal interests supported full scrubbing language in the Clean Air Act Amendments in 1977, the alliance dissolved during the course of the NSPS development over the emission ceiling. A 0.55 lb. ceiling was pushed by air quality groups, but opposed by coal interests as too restrictive on high-sulfur coal reserves.

Had Congress given EPA more flexibility in setting a NSPS for coal-fired power plants by not mandating a percentage reduction of SO_2, the regulatory agency could have set just an emission limit. Such a standard with a low emission limit would have been more cost-effective, have allowed the substitution of low-sulfur coal use for some portion of technological controls, and been easier to enforce.

The final NSPS reduces the demand for low-sulfur coals from the previously existing standard, as desired by high-sulfur coal producers, but does not eliminate it. Lower-sulfur coals will still permit cheaper dry scrubbing, are cleaner to handle and use, and allow a greater margin of safety against violations for utilities that wet scrub.

In one other way Congress does not give EPA the flexibility to attain air quality standards in the most cost-effective manner. The Agency does not have authority to compare the costs and impacts of reducing emissions from existing smelters or existing power plants with the costs and impacts of reducing emissions from new coal-fired power plants. The Agency must set separate, categorical limitations for each of these sources. As this case shows, the great majority of Western sulfur oxide emissions come from inadequately controlled smelters as compared with expected emissions from future power plants; smelters have a special protection in the Clean Air Act from regulations that affect other air pollution sources. Congress treats existing sources of air pollution differently from new sources, because retrofitting pollution controls onto existing plants is usually more costly, technologically more complicated, and politically more difficult than regulating new sources.

THE STANDARD FORCES TECHNOLOGY DEVELOPMENT

The NSPS is technology-forcing in that it encourages:

- Dry scrubbing on low-sulfur coals to an extent not used previously. 70% efficiency for dry scrubbing is the best efficiency currently possible, and it is a developing technology with substantial cost-saving potential;
- Wet scrubbing on high-sulfur coals. Wet scrubbing has worked better on low-sulfur coals, but 90 percent removal efficiencies on high sulfur coals on a 30-day average (85 percent with coal washing credits) have not often been regularly achieved in the United States. By 1987 when the first power plants come on line subject to these new standards, the technology is expected to be much improved over that today;
- Baghouses for particulate control on large utility boilers, required by the 0.03 lb./mill. Btu standard, have not been used before in combination with FGD systems. EPA extrapolated from demonstrated components to show that the required system would perform adequately.

THE ANALYTICAL PROCESS

The NSPS process included the most extensive national impacts analysis EPA has performed for a new standard. EPA's own analysis, in conjunction with DOE, plus the process in which private industry, regulatory reformers and environmentalists, as well as other administration spokesmen, were given full hearing, overcame the initial presumption in favor of full scrubbing. Whereas a model plant analysis showed that more technological controls would reduce emissions more, favoring uniform full scrubbing, the computer modeling of the entire utility industry showed the phenomena of older plants being used longer as new plant costs rise with full scrubbing. This computer modeling showed that somewhat less scrubbing means lower emissions, favoring variable, partial scrubbing. When the theory of uniform reduction was examined with actual projection of impacts, the reality was not what we expected. The initial "hard line" environmental position turned out to be less favorable for the environment, in that partial scrubbing was projected to produce less emissions than full scrubbing. To be sure, environmental groups continued to distrust the modeling exercise, its assumptions, and its conclusions.

As Barber put it, "The standard improved with time."[137] Dry scrubbing, for example, was only fully developed as a feature of the standard in the last few months of the NSPS process, even though it had been discussed earlier. Ultimately, all key EPA officials became satisfied with the NSPS conclusion, whereas they had strongly disagreed earlier in the process. The analytical process combined with broad public participation and EPA agency-wide staffing greatly improved the final product.

The national impacts analyses helped convince people like Douglas Costle and David Hawkins that the initial full scrubbing assumption was not the best for the environment and was unreasonable for industry and energy interests. Costle, Hawkins, and Barber's interest in technical analysis, use of results, and requirement that various agencies concur in factual findings helped make a successful analytical process.

Analyses pointed to what was the most cost-effective standard and helped EPA perform its balancing act with specific quantitative understanding of tradeoffs.

The modeling, for example, helped quantify the regional implications of the standard, showed the impacts in the West of emissions and of demand for coals as compared to the Midwest and East. The modeling helped define the regional policies of the NSPS case. Modeling showed that while the national emissions decrease with the final standard, it means an additional 200,000 tons of SO_2 in the West. However, EPA leaders hope that application of the PSD rule in the West will require greater than NSPS emission limits and cut this impact on a case-by-case basis.

EPA did not perform benefits analysis of the proposed alternatives. To go beyond its cost-effectiveness analysis to cost-benefit analysis would have required the agency to predict the actual locations and sizes of new power plants, then conduct dispersion modeling for each plant to measure the impacts on actual ambient air quality. Next the population exposed to those ambient conditions would need to be estimated along with dose-response studies to determine human impacts of the pollution. EPA had neither the time nor the funds to conduct these studies to support its rule-making, although, to be sure, such analyses would have helped the Agency better understand its regulatory choices. The actual impact on human beings is the primary concern of EPA, not merely emission outputs, although the latter must be used as a substitute impact when data are unavailable.

COSTS OF SETTING THE STANDARD: IT WAS WORTH IT

The management decision for EPA in conducting the NSPS process was whether to invest the funds in the technical analysis, the staff time, and most importantly, top mangement time to support this decision. Costle, Hawkins, Barber, Drayton, and Gamse spent considerable time on this decision from an early point in the process. This early and continuous involvement of top management made this process unique and contributed to its success. But time and funds devoted to the NSPS could not then be devoted to other EPA decisions, and the many other standards EPA must set.

The government investment was about $3 million as shown in Table 8-1. It seems a small investment compared to the $35 billion investment that the utilities must make by 1995 to comply with the standard. Also, possible impacts on air quality, coal companies and coal miners, as well as the political consequences, were high, justifying the investment.

EPA has typically underinvested in technical analysis accompanying its standards setting, usually for lack of funds and mangement time. In this case the Agency sought to raise this one-time front-end investment in better regulations. This is particularly needed in times of economic and energy pressures and the influence of regulatory reformers. This case demonstrates the value of such an investment.

RULE-WRITING PROCESS WITHIN EPA

The regulation development process at EPA was based on joint office participation, which strengthened its conclusions. The OAQPS/OPE joint staffing and funding of contractors broadened the perspective of the standard. OAQPS was the lead unit with engineering and air quality expertise, which was linked on a daily basis with the economists and policy analysts of OPE. ORD, with its engineering expertise and research knowledge, was particularly helpful with dry scrubbing technology and the ORD coal model was used by OAQPS to analyze high-sulfur coal impacts of alternative ceilings. The enforcement division, with its knowledge of implementing a standard on a day-to-day basis, was instrumental is assuring enforceable continuous monitoring requirements and in resolving the issue of averaging times. This was an effective Agency-wide process, although not the formal, established regulation-development process used for other less complicated, less politically volatile standards. Barber took personal charge of the rule-writing, whereas usually it is handled by OAQPS staff. He worked directly with Costle on a regular and frequent basis, so that Costle was

Table 8.1. Cost of Setting NSPS for Fossil-Fuel-Fired Utility Boilers

Costs	*Office*	*Person Years*		*Dollars Expended*
EPA				
Personnel[a]	OAQPS[b]	11.3		$ 575,000
	OPE	1.8		90,000
	ORD	.2		10,000
	Top Management[c]	.5		25,000
		13.8		$ 690,000
Contractor costs	OAQPS			$1,718,000
	OPE			392,000
	ORD			110,000
				$2,220,000
			EPA total	$2,910,000
DOE				
Contractor costs				$180,000
Computer time				35,000
Staff time		1.0		50,000
			DOE total	$265,000
			U.S. Government total	$3,175,000

[a]This is a conservative estimate of personnel costs based on an annual salary for each person year of $25,000, plus $25,000 overhead for each person year.

[b]Includes Walter Barber, not Hawkins or his staff.

[c]Includes an estimate of time for assistant administrators, deputy assistant administrators, and the administrator and his office.

deeply involved in all key decisions and not just presented with a package for approval in a day or two, which often occurs. The working groups, which had initially been established, was disbanded after the joint OAQPS/OPE staffing was arranged. The steering committee review of the draft standard was bypassed, as time expired at the end of the process. The "Red Border" review by assistant administrators was performed, with OAQPS staff walking the final product by the necessary officers in one afternoon. Of course, these officials had been informed by their staffs regularly during standards-development. However, because review periods were short-circuited as court deadlines loomed, some EPA divisions felt that their views were not adequately solicited. Nonetheless, on balance, they concur in the final product.

The rule-development process shows that EPA is no monolith with all like-minded environmentalists promoting controls at any cost, nor all engineers stressing the limits of pollution control technologies. Rather, the Agency includes environmental lawyers, economists, engineers, policy analysts, and practical politicians who represent various, sometimes competitive, forces. When these are linked in a cooperative, institutionalized manner, the interaction can stimulate more complete staff analysis, and better regulatory products than any one element acting alone.

THE INTERAGENCY PROCESS

The interagency process worked well, in large part because friendly, open working relations were established by EPA at the early stage, with DOE, CEA, COWPS, and White House staff. In essence, the joint EPA/DOE analysis in which staff and funding were shared contributed to the kind of trust at the staff level and agreement on facts that supported EPA in the final decision days, or at least moderated opposition. It came about because EPA and DOE leaders mandated their organizations to agree on technical analysis and pursue only policy conclusions at the end of the process.

Costle's briefings and phone calls to top DOE and White House officials built the same kind of confidence at the top policy level.

Cynics on one side of the issue call this a process by which EPA coopted its opponents, but by whatever name, trust and credibility were established through task forces, regular meetings, phone calls and briefings—and these attitudes were critical.

Environmental critics, with the other point of view, see these cooperative interagency steps as giving an advantage to DOE, the White House economists, and regulatory reformers and not the other way around. The other federal units were nothing more than conduits for the regulated interests, environmentalists contend.

It is fair to say, however, that the interagency technical and political process helped produce the evidence and the thinking at EPA that overcame the presumption in favor of full scrubbing. Whether EPA would have changed its collective mind without that interagency process, one can only speculate.

While the White House and executive office of the president were included in the decision process, ultimately the standard was EPA's own, without White House mandates.

One of the fortuitous circumstances of this case was that DOE supported the use of the ICF macroeconomic model, since DOE had sponsored its development and used it previously. This eliminated debate between EPA and DOE during the critical final period of the standard-setting process over the analytical tool.

The macroeconomic models used by EPA, ICF's, UARG's, and NERA's were in basic accord. While aggregate impacts differed in the models, depending on assumptions (oil price, transportation costs, increase in electricity demand), the relative positions of the alternatives, which favored the partial over full scrubbing, were compatible. The one assumption that would have changed the relative cost-effectiveness of alternatives, making full scrubbing preferred to partial, was the cost of scrubbers. If the modeling assumptions about cost are wrong, the conclusion will also be wrong. The model also oversimplifield the utility decision process, by assuming that only least-cost calculations will prevail. Other political and institutional characteristics, or perhaps new federal laws, may alter that least-cost behavior. Also, the macroeconomic analysis of the model did not provide sufficiently specific information about impacts on high-sulfur coal markets. EPA had to supplement the ICF model with its own microeconomic analysis of coal markets.

Thus, while the model was imperfect, it was a useful decision-making tool. Its conclusions had to be combined with other EPA analyses and data and facts and positions presented through the political participation process to produce a realistic conclusion.

IMPORTANCE OF PERSONALITIES AND REGULATORY AGENCY APPOINTMENTS

This case demonstrates the key role of personalities and personal credibility in the human business of writing regulations. Environmental policy-making, like any other fundamental government decision, is based on negotiated solutions, at least in part, within legally established boundaries. Analytical conclusions are interpreted and impacted by personalities and political bargaining. Opinions and preferences of the presidentially-appointed decision-makers are critical elements to the final decision.

A special and helpful circumstance in this case was the fact that many of the principal staffers had either worked together previously or were otherwise compatible personalities, which led to cooperative resolution of various agencies' positions on factual matters.

Barber and Gamse, previously colleagues in OPE, were able to resolve tensions between their two offices (OAQPS and OPE) within EPA, and each office contributed money and regular staff to the analytical effort. Analysts Crenshaw and Haines from OAQPS and Shaver from OPE worked compatibly with DOE staffer Badger. Badger worked for Jim Speyer, who had directed OPE while at EPA in the last administration and transferred to DOE with Al Alm when he became assistant secretary at DOE. Alm himself had a deep understanding of the workings and objectives of EPA, having been its assistant administrator for planning and management. White House staffer Josh Gottbaum had also been on Speyer's staff at DOE and friendly with Badger, Crenshaw, and Haines, prior to going to the White House to be an energy analyst for Alfred Kahn. Gottbaum was the staff person who did the checking of dry scrubbing for Eisenstat during the final decision days and was supportive of EPA's regulatory conclusions.

At the policy level, Costle's credibility within the administration and with various environmental and industry groups, as an environmentally sensitive but politically savvy and fair-minded decision-maker, was crucial to EPA's gaining approval of its preferred standard.

Barber was considered an open and objective analyst in reviewing alternative NSPS, which led various group representatives to believe their views were weighed. His ready access and easy communication with Costle ensured that the administrator had maximum impact on the NSPS decision.

Hawkins' dogged and effective support for the environmentalist full scrubbing standard helped convince environmental groups that their views were being adequately considered within EPA, at least until the end of the process.

EPA's image, or credibility, as an objective, open decision-maker suffered at the end of the NSPS process, as decisions were made quickly under the pressures of the court-ordered deadlines and without the full review and participation of environmental groups that had occurred earlier. EPA staff did not have time to distribute daily revisions to its draft standard and modeling results, as it had previously done. Also, Capitol Hill and White House meetings convinced environmental groups and the press that EPA had "caved in."

Another element that stacked EPA's bargaining process and created an image problem for the Agency was that EPA began the NSPS process with a working assumption for full scrubbing, based on the Clean Air Act and the National Energy Plan debate. This presumption, overcome with analysis and public participation, created the impression of "backing off" from a stricter standard. Also, contributing to this impression was EPA's going public to NAPCTAC initially with the most stringent draft standard that it could defend, assuming that the conclusion after review would reduce the stringency. But having once publicly advocated the extreme engineering position, negotiating a more moderate conclusion was called a "cave in" by environmentalists or a "balance" by coal and utility parties. As it turned out, additional modeling analyses would have pointed to another preferred standard and had EPA waited to go public after that analysis, the Agency's image might have been improved.

In the negotiating process called public participation, industry and environmental groups also advocate extreme positions figuring that they, too, can compromise slightly and still achieve a satisfactory conclusion. Other government agencies do the same. Throughout the bargaining process, the EPA administrator is the judge and policy maker, who must weigh the technical and political evidence and conclude the best policy, while balancing on a tightrope.

LESSONS FOR FUTURE ENVIRONMENTAL RULE-MAKING

This case of environmental rule-writing—for the most part a successful process—provides some lessons for future regulators at EPA and other federal regulatory agencies. From the successes and shortcomings of this NSPS case come insights into the conduct of regulatory analyses, as well as the appropriate roles for federal officials, Congress, and private parties in rule-making.

The Environmental Protection Agency is required by the laws which it administers to set many different kinds of regulations each year. These regulations range from New Source Performance Standards for each industry that emits major amounts of air pollutants, to hazardous waste rules, to national standards governing ambient levels of air and water pollutants, to drinking water standards and many more.

Each of these rules and those written by other federal regulatory agencies will be subject to increased scrutiny as the cost and complexity of federal regulations increase throughout the economy. Presidential and congressional demands are growing for regulatory agencies to assess the potential impacts of proposed rules prior to decision-making and select those alternatives where the benefits justify the costs.

President Ronald Reagan in an executive order issued February 17, 1981 increased the emphasis on regulatory analysis begun by presidents Gerald Ford and Jimmy Carter. The courts have also directed federal regulatory agencies to more thoroughly weigh the costs of their rules and relate these to effectiveness for safety or health. Various bills pending in Congress call for differing types of regulatory analysis prior to regulatory agency decision-making.

This case provides these lessons about regulatory analysis:

- Regulatory analysis is one useful tool to aid regulatory decision-making, as the cost-effectiveness analysis in this case demonstrated. Formal regulatory analyses—cost-benefit studies, cost-effectiveness analyses and risk assessments—should be honed as decision-making tools and more effectively used in regulatory decision-making. At the same time the limits of formal regulatory analyses must be understood. Cost-effectiveness analyses, used by EPA in many of its standards setting, should be used by that agency and other regulators more often. Furthermore, the methodology and data base needed to go beyond cost-effectiveness analyses to conduct cost-benefit analyses should be developed by EPA. The estimation of potential benefits of a regulation needs particular attention;
- Those who expect EPA to analyze alternative regulations should recognize that the number of alternatives and the thoroughness of the analysis depends on available funds. The Office of Management and Budget and the Environmental Protection Agency should assign a higher priority to such analyses than in the past and accordingly increase funds and staff available to support the development of regulations. EPA managers need to allocate the analysis funds according to explicit priorities, so that the greatest amounts support those regulations that will have the greatest impact on the economy, particular geographic areas or population groups, and in those instances where the supporting technical data are weakest;
- The regulatory analysis process needs to be well integrated with the other research and public involvement processes that build to a regulatory decision within each federal regulatory agency, so that the regulation is based on the analyses. While such integration occurred in this case, in many other instances the technical studies do not significantly influence decision-making and may even be prepared after the critical decisions have been made, simply to meet administrative requirements for impact analyses. The studies may not be relevant or timely, or regulatory leaders may not be able or willing to use them. To avoid this situation:

  EPA leaders should make it clear to technical staff that they support regulatory analysis, want matters concerning models and modeling assumptions resolved with other agencies prior to final review of the regulation, want to be included at key junctures of analysis, and will use the results;

  Managers of the regulatory analysis process must understand the factors important to the EPA administrator and incorporate these into the process. Selection of alternatives for study and key assumptions should be presented to the Administrator and Assistant Administrators. As this case shows, some assumptions can determine the preference of one regulatory alternative;

Analysts must be selected who are able to communicate with political leaders of the Agency, avoiding highly detailed, voluminous, and arcane reports. Reports must be timely;

- The regulatory agency should begin regulatory analysis early in the preparation of a rule, withholding the selection of a preliminary position on the rule under development until sufficient findings are available. This will strengthen the content of the draft rule and avoid the necessity for the Agency to repeatedly alter its public position.

- The staffing of regulatory analysis for controversial, major rule-making is stronger if personnel come from various affected divisions within the regulatory agency. Interagency staffing of analysis is appropriate when other departments have a vital stake in the outcome of the rule-making. Such joint staffing can help prevent technical arguments when the final policy decision is debated within the government.

- Regulatory decision-makers need to be aware of the limits of formal regulatory analysis through mathematical modeling. These include methodological and data inadequacies. A model necessarily simplifies reality, data on costs and effects may be lacking, and certain assumptions can predetermine the outcome of analysis.

- If the regulation is based in a significant way on the analysis, peer groups, interest groups, and other affected governmental agencies should be involved in a review and provided an opportunity to comment about basic modeling choices, regulatory alternatives and modeling assumptions. An open rule-writing process is essential, with decisions based on the formal record and ex parte communications avoided.

- EPA should strengthen and increasingly rely on its Agency-wide regulation development process, so that various program offices, enforcement, research, and policy-planning groups contribute to the rule's development. While such a process exists within the Agency, it is not always accorded sufficient weight. In fact, in this case it was bypassed but replaced with a unique agency-wide process.

- This case involved a question EPA often faced: Are we regulating the right pollutants? SO_2 was the pollutant analyzed and regulated in this case, since the national ambient air quality standard is a SO_2 standard. Yet scientists now believe that sulfates—secondary pollutants generated in the atmosphere—are the most damaging to public health, fish, and plant life, although information is insufficient to warrant regulation. The Agency needs to devote more research to the generation and behavior of sulfur oxide and nitrogen oxide pollutants in the atmosphere, their generation, transport, and health and environmental effects.

- This case shows that leaders of regulatory agencies need an appreciation and understanding of regulatory analyses—its scientific, economic, and policy-analysis aspects—a judicial temperament that can seek and balance many different points of view that come from within and outside the Agency, environmental awareness, and the political judgment to strike the proper public policy balance of national objectives.

- If Congress wants a regulatory agency to choose cost-effective regulations, it must write the laws that mandate the rules in a manner sufficiently general to allow EPA analysis and choice. In this case, the Clean Air Act required a percentage reduction figure, as well as an emission limit, which prevented the Agency from selecting the most cost-effective standard, which would have been a low emission limit and no percentage reduction.

EPILOGUE

On April 29, 1981 the U.S. Court of Appeals affirmed EPA's rule for air pollution controls on utilities' new coal-fired steam generators. In a 253-page decision, the Court upheld EPA's rule as "reasonable" on both substantive and procedural grounds. The court concluded that the rule furthered the policies of the Clean Air Act, and that there was sufficient documentation in the record for a variable standard, reliance on dry scrubbing, and the percent reduction figures for sulfur dioxide and particulate matter. The Court concluded that the standard encourages technological innovation in pollution controls and that procedures provided adequate notice to the public, consideration of various public views, and the appropriate relations with Congress, other executive agencies, the White House staff, and the President.

The Court had the following remarks about the key regulatory issues:

Variable percentage reduction option: "EPA has the authority under section 111 of the Act to promulgate the variable standard . . . "[138]"The results of EPA's econometric modeling" which forecast substantial benefits to be obtained by adopting the variable standard, provide adequate support for EPA's decision . . . EPA's regulatory analysis also persuade us that the variable standard does indeed advance policies of the Act . . . "[139]

Regulatory analysis: "EPA was justified in relying on long-term analysis of national and regional cost, environmental and energy impacts of alternative percentage reduction standards in order to select the 'best technological system' upon which to base the NSPS. . . . We are more sympathetic to Sierra Club's complaint about the reliability of EPA's econometric model. Such models, despite their complex design and aura of scientific validity, are at best imperfect and subject to manipulation as EPA forthrightly recognizes . . . Still we cannot agree with Sierra Club that it was improper for EPA to employ an econometric computer model, or hold as a matter of law that EPA erred by relying on the model to forecast the future impacts of alternative standards fifteen years hence . . . We conclude that EPA's reliance on its model did not exceed the bounds of its usefulness and that its conduct of the modeling exercise was proper in all respects."[140]

Dry scrubbing: "EPA's considerations of dry scrubbing as a reason for its selection of a nonuniform standard is consistent with the agency's authority under section 111 and that there is support in the record for doing so . . . Support in the record for selecting 70 percent as a magic percentage for encouragement of dry scrubbing is less than overwhelming . . . However, it was reasonable for EPA to seek to encourage dry scrubbing and to be concerned with the effect of the NSPS on the future of new technology."[141]

Procedures in setting variable standard: While not completely in accord with EPA's procedures in setting the 70 to 90 percent variable standard for sulfur dioxides, the Court said, "This rulemaking was by no means a neat and tidy proceeding, and it might well have been the wiser course if EPA had chosen to publish a new proposal for another round of comment, but we cannot say that the absence of new notice and comment is a fatal defect . . . we decline to remand the variable standard to EPA on the procedural grounds argued by Sierra Club."[142]

90 percent removal standard for sulfur dioxide: The Court rejected the utility industry's argument that the 90 percent reduction of potential sulfur dioxide emissions was unachievable, saying "EPA adequately demonstrated the achievability of the 90 percent standard"[143]

Particulate standard: The Court upheld the particulate standard against the utility industry challenge, saying "EPA has sufficiently established that the standard for particulate emissions is achievable by baghouse control by accounting for relevant variables and by demonstrating the representativeness of its data. . . . We affirm the particulate standard on the basis of the performance of baghouse technology."[144]

1.2 lb/MBtu emission ceiling and "ex parte" contacts: The Court concluded that "EPA's adoption of the 1.2 lbs./MBtu emissions ceiling was free from procedural error. The post-comment period contacts here violated neither the statute nor the integrity of the proceeding . . . it was not improper for the agency to docket and consider documents submitted to it during the post-comment period, since no document vital to EPA's support for the rule was submitted so late as to preclude any effective comment."[145]

As to meetings with Senator Byrd, the Court stated, "Administrative agencies are expected to balance Congressional pressure with the pressures emanating from all other sources."[146] Regarding White House contacts, the Court stated, "Unless expressly forbidden by Congress, such intra-executive contacts may take place, both during and after the public comment period; the only real issue is whether they must be noted and summarized in the docket."[147]

In its conclusion, the Court of Appeals aptly summarized the history and possible future of the NSPS for coal-fired power plants as well as the complexity of EPA's task in setting such a rule:

> Since the issues in this proceeding were joined in 1973 when the Navajo Indians first complained about sulfur dioxide fumes over their Southwest homes, we have had lawsuits, almost four years of substantive and procedural maneuvering before the EPA, and now this extended court challenge. In the interim, Congress has amended the Clean Air Act once and may be ready to do so again. The standard we uphold has already been in effect for almost two years, and could be revised within another two years.
>
> We reach our decision after interminable record searching (and considerable soul searching). We have read the record with as hard a look as mortal judges can probably give its thousands of pages. We have adopted a simple and straight-forward standard of review, probed the agency's rationale, studied its references (and those of appellants), have endeavored to understand them where they were intelligible (parts were simply impenetrable), and on close questions given the agency the benefit of the doubt out of deference for the terrible complexity of its job. We are not engineers, computer modelers, economists or statisticians, although many of the documents in this record require such expertise—and more.
>
> Cases like this highlight the enormous responsibilities Congress has entrusted to the courts in proceedings of such length, complexity and disorder. Conflicting interests play fiercely for enormous stakes, advocates are prolific and agile, obfuscation runs high, common sense correspondingly low, the public intent is often obscured.
>
> We cannot redo the agency's job; Congress has told us, at least in proceedings under this Act, that it will not brook reversal for small procedural errors; *Vermont Yankee* reinforces the admonition. So in the end we can only make our best effort to understand, to see if the result makes sense, and to assure that nothing unlawful or irrational has taken place. In this case, we have taken a long while to come to a short conclusion: the rule is reasonable.[148]

NOTES

1. U.S. Environmental Protection Agency, "EPA Tightens Air Pollution standards for New Power Plants," Press release dated May 25, 1979, p. 1.
2. Ibid.
3. *Environment Reporter*, Bureau of National Affairs, June 1, 1979.
4. *Wall Street Journal*, May 29, 1979, p. 12.
5. *Washington Post*, June 3, 1979.
6. "EPA Tightens Air Pollution Standards," op. cit., p. 2.
7. Author's interview with Walter Barber, Deputy Assistant Administrator for Air Quality Planning and Standards, EPA, July 16, 1979.
8. 36 *Federal Register* 24, 876-&& (1971), 40 C.F.R. 60.40 (1977).
9. Daniel B. Badger, Jr., "New Source Standard for Power Plants I: Consider the Costs," *Harvard Environmental Law Review*, Vol. 3:48, 1979, p. 50.
10. Letter from Carl E. Bagge, President, National Coal Association, to Douglas M. Costle, Administrator, U.S. Environmental Protection Agency, April 23, 1979.
11. Oljato Chapter of the Navaho Tribe v. Train, 515 F. 2d 654 (D.C. Cir. 1975).
12. E.P.A. Docket No. 11-D-24.
13. Section 111(–) (1) (C), H.R. 10498, 94th Congress, 2nd Session (1976).
14. H. Rept. No. 1175, 94th Congress, 2nd Session (1976).
15. H.R. 10498, 94th Congress, 2nd Session.
16. S. 3219, 94th Congress, 2nd Session.
17. Conference Report on S. 3219, 94th Congress, 2nd Session (1976).
18. David P. Rall et al., Report of the Committee on Health and Environmental Impacts of Increased Coal Utilization, Dec. 1977, reprinted in 43 *Federal Register* 2229, 2231, 1978.
19. Executive Office of the President, the National Energy Plan, 1977, Fact Sheet, April, 1977.
20. H. Rep. No. 294, 95th Congress, 1st Session (1977), p. 187.
21. Sec. 125 (b) (3), Public Law. No. 95-95, 91 Stat. 685 (1977).
22. H.R. 6161, 95th Congress, 1st Sesstion, Sec. 111 (a) (1) (a) (1977); 42 U.S.C.A. Sec. 7411 (a) (1) (Supp. 1978).
23. The Utilities Air Regulatory Group is an ad hoc group formed for the purpose of representing the utility industry's interests in air quality legislation and rule-making.

The organization consists of the Edison Electric Institute, National Rural Electric Cooperatives Association, and 63 independent utility systems.

24. Conf. Report. H.R. Rep. No. 564, 95th Congress, 1st Session (1977).
25. Ibid.
26. "EPA Tightens Air Pollution Standards, " op. cit., p. 3.
27. Ibid.
28. Ibid.
29. Lester Lave and Eugene Suskin, *Air Pollution and Human Health*, Baltimore: Johns Hopkins University Press for Resources for the Future, 1977, p. 224–225.
30. Environmental Protection Agency, "New Stationary Sources Performance Standards, Electric Utility Steam Generating Units," *Federal Register*, Vol. 44, No.113 (June 11, 1979), p. 33606.
31. Letter of R. W. Scherer, President, Georgia Power Company, to Jimmy Carter, President of the United States, February 3, 1978.
32. Richard E. Ayers, et al., letter to President Jimmy Carter, August 3, 1978.
33. Memorandum from David G. Hawkins, Assistant Administrator for Air and Waste Management, EPA, to Douglas M. Costle, Administrator, EPA, Subject: "Revised NSPS for Power Plants, " November 21, 1977.
34. Ibid.
35. Memorandum from Thomas F. Schrader, Energy Policy Staff, OPE, to Jack Farmer, Chief, Standards Development Branch, ESED, Subject: "Working Group Meeting Notes—Revision of NSPS for Electric Utility Boilers," November 9, 1977.
36. Memorandum from Roy Gamse, Deputy Assistant Administrator for Planning and Evaluation, to Walter Barber, Deputy Assistant Administrator for Air Quality Planning and Standards, January 18, 1978.
37. Ibid.
38. Memorandum from Roy Gamse to Walter Barber, February 28, 1978.
39. Memorandum from Don Goodwin, Director, Emission Standards and Engineering Division, to Walter Barber, January 19, 1978.
40. Memorandum from George W. Walsh to Don Goodwin, Subject: "Revising the SO_2 NSPS for Steam Generators," April 21, 1976.
41. Minutes of the Meeting of the National Air Pollution Control Techniques Advisory Committee, December 14 and 15, 1977, Alexandria, Virginia.
42. Ibid.
43. Letter from Joseph J. Brecher to Don Goodwin, for the Record of the NAPCTAC Meeting, December 31, 1977.
44. Ibid.
45. Letter of R.S. Scherer, President, Georgia Power Company, to Jimmy Carter, President of the United States, February 3, 1978.
46. Letter of W. Robert Worley, Georgia Power Company, to Hon. Herman E. Talmadge, senator from Georgia, February 3, 1978.
47. "Memorandum for Ambassador Robert Strauss from the Business Roundtable, New Source Performance Standards Under the Clean Air Act Amendments," June 22, 1978.
48. Letter from James M. Evans to Hon. Stuart E. Eizenstat, July 12, 1978.
49. Author's interview with Walter Barber, July 16, 1979.
50. Speech of John O'Leary, November 7, 1977.
51. Testimony of Dr. James Liverman, Acting Assistant Secretary for Environment, Department of Energy, before the House Science and Technology Committee, April 26, 1978.
52. Maxwell, Elder and Morasky, "Sulfur Oxides Control Technology in Japan," Report

to Senator Henry M. Jackson, Chairman, U.S. Senate Committee on Energy and Natural Resources, June 30, 1978, p. 20.

53. Letter from Senators Henry M. Jackson and Clifford Hansen to Jimmy Carter, President of the United States, June 23, 1978.
54. Letter from Senator Henry M. Jackson, et al., to Douglas M. Costle, June 27, 1978.
55. Letter from Scott M. Matheson, Governor of Utah, to Douglas M. Costle, June 19, 1978.
56. Memorandum from Roy Gamse to Walter Barber, February 28, 1978, Subject: "Outstanding Issues on the NSPS for Electric Utility Boilers."
57. Memorandum from David G. Hawkins to Deputy Administrator Barbara Blum, Subject: "Control of Sulfur Oxides Emissions for New Coal Fired Power Plants," June 28, 1978.
58. Interview of the author with John D. Crenshaw, August, 1980. *See also* memorandum from John D. Crenshaw, Program Analyst, Energy Information Section, ESB to Don Goodwin, Director, Emission Standards and Engineering Division, Subject: "SO_2 Emission Assumptions for the Utility NSPS, Differences in ICF and Teknekron Models," September 13, 1978.
59. Ibid.
60. ICF, Inc., *Further Analysis of Alternative NSPS for New Coal-Fired Power Plants, Preliminary Draft,* September, 1978, p. 3.
61. Ibid.
62. Ibid, p. 4.
63. Ibid.
64. UARG Briefing Paper for Meeting with EPA Administrator Costle, on August 30, 1978, on Revised New Source Performance Standards for Fossil Fired Electric Generating Plants, pp. 5–6.
65. Ibid.
66. Memorandum from Dave Shaver, Policy Planning Division, OPE, to William Drayton, Assistant Administrator for Planning and Management, Subject: "Power Plant Delays Due to BACT Determination with Partial Control NSPS." August 16, 1978.
67. Letter from Robert Ayres, to Douglas M. Costle, May 1, 1978.
68. *New York Times,* June 12, 1978, p. 1.
69. Ibid.
70. The Sierra Club filed a complaint on July 14, 1978, with the U.S. District Court for the District of Columbia requesting injunctive relief to require, among other things, that EPA propose the revised standards by August 7, 1978. A consent order was developed and issued by the court requiring the EPA Administrator to deliver the proposal package to the office of the *Federal Register* by September 12 1978 and promulgate the final standards within 6 months after proposal.
71. Memorandum from Bill Nordhaus to Bill Drayton, Subject: "Benefits of Alternative NSPS for SO_2 Emissions from Steam Electric Power Plants," July 11, 1978.
72. Author's interview with Walter Barber, July 16, 1979. *See also* note from Bill Hamilton to Walter Barber penned on a memorandum dated July 21, 1978.
73. Letter from John F. O'Leary, Deputy Secretary, Department of Energy, to Douglas M. Costle, August 11, 1978.
74. Memorandum from Douglas M. Costle to Stuart Eizenstat, August 9, 1978.
75. Ibid.
76. Author's Interview with David Hawkins, July 2, 1979.
77. Environmental Protection Agency, "Electric Utility Steam Generating Units: Proposed Standards of Performance and Announcement of Public Hearing on Proposed

Standards," *Federal Register*, Vol. 43, No. 182, Tuesday, September 19, 1978, Part V., p. 42154.

78. Ibid.
79. Richard E. Ayres, David Doniger, Veronica M. Kun for Natural Resources Defense Council and Robert J. Rauch, David Lennett for Environmental Defense Fund, *Comments on Proposed Standards of Performance for New Electric Utility Steam Generating Units*, January 15, 1979, p. IV-7.
80. Harmon Wong-Woo and Alan Goodley, California Air Resources Board, et al., "Observation of Flue Gas Desulfurization and Denitrification Systems in Japan," March 7, 1978, p. 6.
81. Daniel B. Badger, Jr., op. cit., p. 53.
82. Richard Ayres, et al., *Comments on Proposed Standards*, op. cit., p. v-4.
83. Richard E. Ayres and David D. Doniger, "New Source Standard for Power Plants II: Consider the Law," *Harvard Environmental Law Review*, Vol. 3:48, 1979.
84. Daniel B. Badger, Jr., op. cit., p. 55.
85. Ibid.
86. Ibid.
87. Letter from John F. O'Leary, Deputy Secretary of Department of Energy, to Douglas M. Costle, December 15, 1978.
88. UARG Briefing Paper for Meeting with EPA Administrator Costle, on August 30, 1978. *See also* "Comments of the Utility Air Regulatory Group on Proposed Standards of Performance for New Electric Utility Steam Generating Units." December 15, 1978, corrected version January 15, 1979.
89. Report of the Regulatory Analysis Review Group Submitted by the Council on Wage and Price Stability, "Environmental Protection Agency's proposed Revision of NSPS for Electric Utility Steam Generating Units," January 15, 1979, p. 42.
90. Ibid., p. 5.
91. John Haines, Assistant to the Director, ESED, "Full Versus Partial Control," January 29, 1979.
92. Memorandum from David W. Tundermann, to Bill Drayton, Subject: "NSPS Update," March 9, 1979, pp. 2–3.
93. Author's interview with Walter Barber, July 16, 1979.
94. Letter from Carl E. Bagge, President of the National Coal Association, to Douglas M. Costle, April 6, 1979.
95. Memorandum from John Crenshaw to Walter Barber, May 11, 1979.
96. Letters from Carl Bagge to Douglas M. Costle, April 6, April 20, and April 23, 1979.
97. Memorandum of John D. Crenshaw to Walter Barber, Subject: "Emission Ceiling—SO_2 NSPS for Utility Boilers," May 11, 1979.
98. Ibid.
99. Memorandum from David W. Tundermann to Bill Drayton, March 9, 1979, pp. 2–3.
100. Ibid.
101. David Hawkins continued to be concerned about impairment of Western visibility with a partial scrubbing standard. Thus, Barber directed his staff to contact several recognized authorities on visibility to assess likely impacts of full versus partial scrubbing in the West. They concluded that there would be no perceived impacts from either alternative in 1995 and only very slight in 2010. *See* Memorandum from Jack Backman, Strategies and Air Standards Divisions to Walter Barber, Director, OAQPS, Subject: "NSPS Visibility Impact," March 7, 1979.
102. Comments of Walter Barber on author's draft case study, July, 1980.
103. Memorandum of Frank Princiatta to Walter Barber, February 26, 1979.
104. Memorandum from David Shaver to Bill Drayton, April 11, 1979.
105. Author's interview with David Hawkins, July 2, 1979.

106. *Air/Water Pollution Report* April 23, 1979.
107. Memorandum from David Hawkins, to the NSPS Docket, April 24, 1979.
108. *Air/Water Pollution Report*, April 23, 1979.
109. Memorandum from David Hawkins to the Files, Subject: "Power Plants NSPS Meeting with Senator Byrd," May 2, 1979.
110. *Washington Post*, May 5, p. 1.
111. Memorandum from John Crenshaw to John Haines, Subject: "Sulfur Dioxide Emissions Impact of a 1.0 Versus 1.2 Emission Limit," May 30, 1979.
112. Ibid.
113. Author's interview with David Hawkins, July 2, 1979.
114. Environmental Protection Agency, "Standards for Performance for New Stationary Sources for Electric Utility Steam Generating Units; Decision in Response to Petitions for Reconsideration," *Federal Register*, Vol. 45, No. 26, February 6, 1980, p. 8215.
115. Statement of Douglas M. Costle, May 25, 1979, p. 6.
116. *Washington Post*, May 25, 1979, p. 6.
117. Environmental Protection Agency, "New Stationary Sources Performance Standards; Electric Utility Steam Generating Units," *Federal Register*, Vol. 44, No. 113, June 11, 1979.
118. Statement of Douglas M. Costle, May 25, 1979.
119. Author's interview with David Hawkins, July 2, 1979.
120. *Environment Reporter*, June 1, 1979, p. 143.
121. Ibid.
122. *Environment Reporter*, May 11, 1979, p. 35.
123. Ibid.
124. Ibid.
125. Ibid.
126. UARG petitioned EPA to review the entire standard.
127. Environmental Protection Agency, "Standards of Performance for New Stationary Sources for Electric Utility Steam Generating Units; Decision in Response to Petitions for Reconsideration," *Federal Register*, Vol. 45, No. 26 February 6, 1980, pp. 8210–8233.
128. Ibid.
129. Ibid, p. 8214.
130. Ibid. p. 8220.
131. Sierra Club et al. *v*. Douglas M. Costle, Administrator, Environmental Protection Agency No. 79–1565 (D.C. Cir., filed June 11, 1979).
132. Sierra Club et al. *v*. Douglas M. Costle, Administrator, Environmental Protection Agency, No. 79-1565 (D.C. Cir., filed June 11, 1979), Brief for Petitioner, Environmental Defense Fund, p. 21.
133. Ibid.
134. Ibid.
135. Sierra Club et al. *v*. Costle, Brief for Respondents, p. 141.
136. Ibid, p. 143.
137. Author's interview with Walter Barber, June 11, 1980.
138. U.S. Court of Appeals for the District of Columbia, No. 79-1565. Sierra Club *v*. Douglas M. Costle, April 29, 1981, p. 31.
139. Ibid. p. 66.
140. Ibid, p. 52-53, 56.
141. Ibid, p. 95.
142. Ibid, p. 104.
143. Ibid, p. 143.

144. Ibid, p. 166.
145. Ibid, p. 224.
146. Ibid, p. 224
147. Ibid, p. 213.
148. Ibid, p. 225-226.

APPENDIX 1

Results of Early Contractor Studies for the Environmental Protection Agency Identifying Costs and Impacts of Alternative Proposed NSPS for Coal-Fired Power Plants

Table A1.1: Costs of SO_2 Control Alternatives for Level of 1.2 lb./10^6 Btu

Coal Type Percent Sulfur	Lime FGD				Limestone FGD			
Boiler Capacity (megawatts)	Capital $/kW	O&M[a] mill./kWh	Fixed mill./kWh	Total	Capital $/kW	O&M[a] mill./kWh	Fixed mill./kWh	Total
Eastern, 3.5								
25	262.76	10.38	10.50	20.88	278.48	10.78	11.13	21.91
100	234.78	6.83	9.38	16.21	285.77	7.25	11.42	18.67
200	174.46	5.48	6.97	12.45	202.46	5.58	8.09	13.67
500	124.93	4.69	4.99	9.68	142.58	4.45	5.70	10.15
1000	103.71	4.09	4.14	8.23	119.02	4.03	4.76	8.79
Eastern, 7.0								
25	321.06	12.46	12.83	25.29	338.64	12.88	13.53	26.38
100	290.51	8.76	11.61	20.43	352.58	9.01	14.09	23.10
200	217.36	7.21	8.68	15.89	252.11	7.10	10.07	17.17
500	156.34	6.06	6.25	12.31	185.94	5.86	7.43	13.29
1000	131.59	5.56	5.26	10.82	156.53	5.32	6.25	11.57
Anthracite								
500								
Lignite								
500	96.67	3.06	3.86	6.90				

[a]Mills per killowat hour.
Source: PEDCo., 1977a.

Mag-Ox FGD				*Double Alkali FGD*				*Wellman-Lord FGD*			
Capital $/kW	*O&M[a] mill./ kWh*	*Fixed mill./ kWh*	*Total*	*Capital $/kW*	*O&M[a] mill./ kWh*	*Fixed mill./ kWh*	*Total*	*Capital $/kW*	*O&M[a] mill./ kWh*	*Fixed mill./ kWh*	*Total*
409.12	14.52	16.35	30.87	308.00	11.16	12.31	23.47	370.24	13.19	14.79	27.96
319.66	8.59	12.77	21.36	297.70	7.84	11.89	19.73	290.22	7.36	11.60	18.96
210.41	6.17	8.41	14.58	194.78	5.91	7.78	13.69	188.91	5.21	7.55	12.76
156.76	5.10	6.26	11.36	147.09	5.14	5.88	11.02	141.16	4.20	5.64	9.84
131.00	4.43	5.23	9.66	119.60	4.47	4.78	9.25	116.05	3.58	4.64	8.22

Table A1.2: Costs of SO_2 Control Alternatives for 90 Percent SO_2 Removal

Coal Type Percent Sulfur Boiler Capacity (megawatts)	Lime FGD				Limestone FGD			
	Capital $/kW	O&M[a] mill./kWh	Fixed mill./kWh	Total	Capital $/kW	O&M[a] mill./kWh	Fixed mill./kWh	Total
Eastern, 3.5								
25	289.96	11.01	11.59	22.60	307.60	11.43	12.29	23.72
100	259.88	7.44	10.38	17.82	316.67	7.88	12.65	20.53
200	194.23	5.99	7.76	13.75	225.08	6.09	8.99	15.08
500	139.40	5.15	5.57	10.72	100.15	4.90	6.40	11.30
1000	115.91	4.50	4.63	9.13	134.06	4.44	5.36	9.80
Eastern, 7.0								
25	322.48	12.64	12.88	25.52	339.88	13.03	13.58	26.61
100	291.52	8.91	11.65	20.56	353.70	9.16	14.13	23.29
200	218.09	7.35	8.71	16.06	252.70	7.23	10.10	17.33
500	157.17	6.30	6.28	12.58	186.59	6.00	7.46	13.46
1000	132.02	5.69	5.27	10.96	157.05	5.45	6.27	11.72
Western, 0.8								
25	252.52	9.68	10.09	19.77	269.16	10.10	10.75	20.85
200	166.09	4.91	6.64	11.55	192.13	5.18	7.68	12.86
500	119.42	3.95	4.77	8.72	137.20	4.11	5.48	9.59
Anthracite								
500	116.57	3.67	4.66	8.33				
Lignite								
500	116.71	3.71	4.66	8.37				

[a]Mills per kilowatt hour.

Source: PEDCO., 1977a.

Magnesium Oxide FGD				Double Alkali FGD				Wellman-Lord FGD			
Capital $/kW	O&M[a] mill./ kWh	Fixed mill./ kWh	Total	Capital $/kW	O&M[a] mill./ kWh	Fixed mill./ kWh	Total	Capital $/kW	O&M[a] mill./ kWh	Fixed mill./ kWh	Total
439.40	15.15	17.56	32.71	337.60	11.86	13.49	25.35	395.76	13.65	15.81	29.46
354.17	9.26	14.15	23.41	329.64	8.56	13.17	21.73	317.53	7.82	12.69	20.51
233.21	6.68	9.32	16.00	215.57	6.47	8.61	15.08	206.02	5.54	8.23	13.77
174.60	5.59	6.98	12.57	163.79	5.67	6.54	12.21	154.93	4.51	6.19	10.70
146.93	4.84	5.87	10.71	133.76	4.92	5.34	10.26	127.56	3.81	5.10	8.91
481.56	16.56	19.24	35.80	374.92	13.75	19.98	33.73	417.48	14.01	16.68	30.67
398.03	10.37	15.90	26.27	364.41	10.25	14.56	24.81	336.49	8.04	13.44	21.48
264.99	7.81	10.59	18.40	242.43	8.04	9.69	17.73	220.93	5.22	8.83	14.05
205.81	6.77	8.22	14.99	183.80	7.32	7.34	14.66	169.74	4.81	6.76	11.57
173.01	5.85	6.91	12.76	151.11	6.30	6.04	12.34	140.16	4.00	5.61	9.61

Table A1.3: Nationwide Costs of Generating Electricity Under Selected Alternative of NSPS 1986–1995 High Rate of Power Growth

	Baseline[a]	*90 Percent SO_2 Removal*[b]		*80 Percent SO_2 Removal*	
Allocation	*Dollars*[c]	*Dollar cost*[c]	*Increment over baseline (dollars)*	*Dollar cost*[c]	*Increment over baseline (dollars)*
Total revenue	1,346.1[d]	1,387.1[d]	41.0[d]	1,393.6[d]	42.5[d]
Total	1,197.1	1,250.0	52.9	1,251.4	54.3
Fuel	432.6	435.2	2.6	435.3	2.7
Pollution control	40.0	54.9	14.9	51.1	11.1
Other operation and maintenance	196.1	197.0	1.0	196.9	0.8
Pollution control investment	17.6	48.3[b]	30.7	47.7	30.1
Net profit	140.0	137.1	−2.9	137.2	−2.8
All other investments[d]	518.8	524.2	5.4	523.1	4.3

[a]Continuation of present standard—520 ng/J (1.2 lb. $SO_2 10^6$ Btu).
[b]Assuming a particulate level of 0.1 lb/10^6 Btu.
[c]Billions of 1975 dollars.
[d]Excluding pollution control.
[e]Detailed items do not sum to totals because miscellaneous costs were not included.
Source: Teknekron, Inc., 1978.

Table A1.4: Nationwide Costs of Generating Electricity under Alternative NSPS 1986–1995 Moderate Rate of Power Growth

	Baseline[a]	*90 Percent SO_2 Removal*		*80 Percent SO_2 Removal*		*0.5 lb. $SO_2/10^6$ Btu*	
Allocation	*(dollars)*[b]	*Dollar Cost*[b]	*Increment over Baseline*	*Dollar Cost*[b]	*Increment over Baseline*	*Dollar Cost*[b]	*Increment over Baseline*
Total revenue	1,149.5	1,171.6	22.1	1,170.4	20.9	1,170.0	20.5
Total cost	1,023.9	1,050.6	26.7	1,049.7	25.8	1,049.4	25.5
Fuel	369.8	369.5	−0.3	369.5	−0.3	368.9	0.9
Pollution control	36.4	43.6	7.2	43.3	6.9	43.4	7.0
Other operation and maintenance[c]	167.4	167.9	0.5	165.9	0.5	167.9	0.5
Pollution control investment	8.0	21.0	13.0	21.3	13.3	21.3	13.3
All other investments[c]	325.3	329.5	4.2	328.4	3.1	328.7	3.4
Net profit	125.6	121.0	−4.6	120.7	−4.9	120.6	−5.0

[a]Continuation of present standard −520 ng/J (1.2 lb. $SO_2/10^6$ Btu).
[b]Billions of 1975 dollars.
[c]Excluding pollution control.
[d]Ratio of increment to baseline cost (where this is a positive value).
Source: Teknekron, Inc., 1978.

Table A1.5: Energy Penalty in Mills./kWh[b] for Selected Control Processes, Scenarios, and Plant Sizes

			FGD Process Lime or Limestone*		
Coal Type	*Percent Sulfur*	*Plant Capacity in MW*	*1.2 lb. SO_2/ 10^6 Btu*	*90% SO_2 reduction*	*0.5 lb. SO_2/ 10^6 Btu*
Eastern	3.5 to 7.0	25	0.88	0.97	—
		200	0.80	0.88	—
		500	0.77	0.85	—
		1000	0.75	0.83	—
Eastern (with coal cleaning)	3.5	500	12.68	—	—
Anthracite	0.8	500	0.77	0.85	—[b]
Western	0.8	25	—[a]	1.16	0.97
		200		1.06	0.88
		500		1.02	0.85
Western lignite	0.4	500	0.77	0.85	—

*Costs calculations as reported were virtually identical for both processes.
[a]Not reported.
[b]Costs represent the purchase price of power from the unit to run the FGD system.
Source: PEDCo. Environmental Special Communication, January 31, 1978.

and Scenario Magnesium Oxide		Double Alkali		Wellman-Lord	
1.2 lb. SO_2/ 10^6 Btu	90% SO_2 reduction	1.2 lb. SO_2/ 10^6 Btu	90% SO_2 reduction	1.2 lb. SO_2/ 10^6 Btu	90% SO_2 reduction
1.05	1.16	0.88	0.97	1.05	1.16
0.95	1.05	0.80	0.88	0.95	1.05
0.92	1.02	0.77	0.85	0.92	1.02
0.90	0.99	0.75	0.83	0.90	0.99
—	—	—	—	—	—
—	1.02	—	0.85	—	1.02
—	1.39	—	1.16	—	1.39
	1.26		1.06		1.26
	1.22		1.02		1.22
—	1.02	—	0.85	—	1.02

Table A1.6: Incremental Costs of Removing 90 Percent SO_2 (Compared to costs of meeting 1.2 lb./10^6 Btu)

Control Alternative	*Lime FGD*				*Limestone FGD*			
Coal Type, Percent Sulfur, Boiler Capacity (megawatts)*	*Capital $/kW*	*O & M mill./kWh*	*Fixed mill./kWh*	*Total mill./kWh*	*Capital mill./kWh*	*O & M $/kwh*	*Fixed mill./kWh*	*Total mill./kWh*
Eastern, 3.5								
25	27.20	0.63	1.09	1.72	29.12	0.65	1.16	1.81
100	25.10	0.61	1.00	1.61		0.63	1.23	1.86
200	19.77	0.51	0.79	1.30	22.62	0.51	0.90	1.41
500	14.53	0.46	0.58	1.04	17.57	0.45	0.70	1.15
1000	12.20	0.41	0.49	0.90	15.04	0.41	0.60	1.01
Eastern, 7.0								
25	1.40	0.18	0.05	0.23	1.24	0.15	0.05	0.20
100	1.01	0.15	0.04	0.19	1.12	0.15	0.04	0.19
200	0.73	0.14	0.03	0.17	0.65	0.14	0.03	0.16
500	0.83	0.24	0.03	0.27	0.65	0.14	0.03	0.17
1000	0.43	0.13	0.01	0.14	0.52	0.12	0.02	0.14
Lignite								
500	20.04	0.65	0.80	1.45				

*Western 0.8 percent sulfur coal and Eastern anthracite coal are not shown. These are assumed to require no FGD to meet current standards. Therefore, the total costs for removing 90 percent SO_2 are "incremental" over the baseline situation.

Source: PEDCo., 1977a.

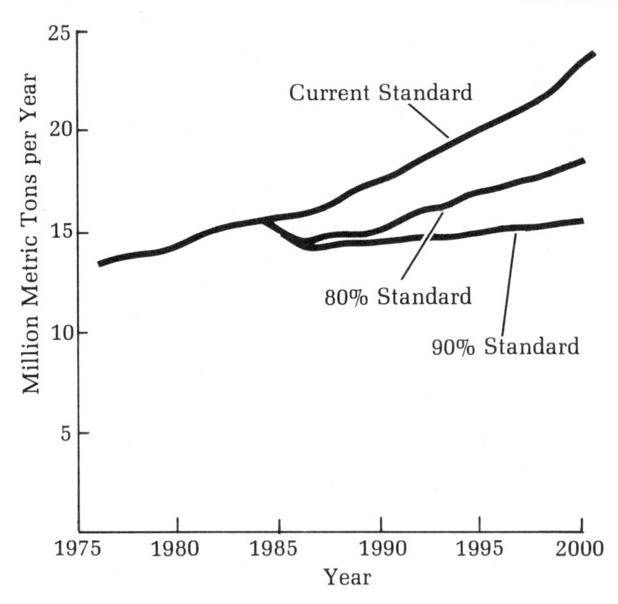

Figure A-1. National power-plant SO_2 emissions under alternative control scenarios, high growth.

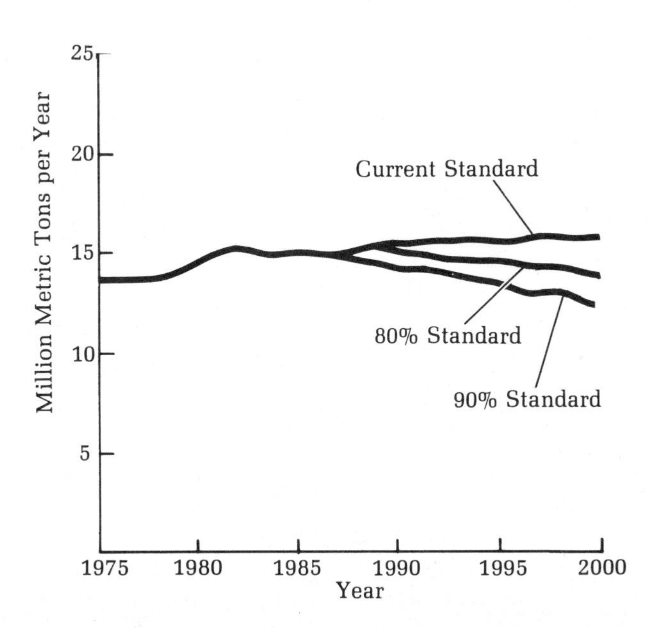

Figure A-2. National power-plant SO_2 emissions under alternative control scenarios, moderate growth.

Table A1.7: Regional and National Power-Plant SO_2 Emissions* Assuming High Growth

Region	*1985*	*1990*	*1995*	*2000*
		Current Standard		
NE	0.25	0.26	0.32	0.44
MA	1.67	1.62	1.72	1.87
SA	4.04	4.19	4.67	5.18
ENC	4.29	4.75	5.45	6.14
ESC	2.43	2.28	2.26	2.44
WNC	1.56	1.63	1.93	2.29
WSC	0.94	1.84	2.71	3.49
NM	0.07	0.14	0.28	0.42
SM	0.23	0.31	0.38	0.56
PA	0.32	0.43	0.60	0.90
National	15.90	17.50	20.30	23.80
		80% Standard		
NE	0.24	0.24	0.29	0.38
MA	0.00	1.67	1.97	2.16
SA	3.75	3.96	4.04	4.09
ENC	3.66	4.11	4.78	5.56
ESC	2.40	2.30	2.25	2.51
WNC	1.30	1.34	1.65	1.91
WSC	0.75	0.98	1.22	1.44
NM	0.06	0.09	0.12	0.15
SM	0.20	0.20	0.21	0.24
PA	0.26	0.26	0.26	0.32
National	14.30	15.20	16.80	18.80
		90% Standard		
NE	0.23	0.20	0.22	0.24
MA	1.65	1.54	1.59	1.60
SA	3.70	3.62	3.46	3.27
ENC	3.60	3.91	4.29	4.67
ESC	2.39	2.27	2.37	2.47
WNC	1.30	1.31	1.59	1.75
WSC	0.73	0.82	0.90	0.98
NM	0.06	0.07	0.08	0.09
SM	0.19	0.17	0.15	0.15
PA	0.25	0.24	0.22	0.24
National	14.10	14.20	14.90	15.50

*Million metric tons per year.

Source: Teknekron, Inc.

Table A1.8: Regional and National Power-Plant SO_2 Emissions Assuming Moderate Growth

Region	*1976*	*1985*	*1990*	*1995*	*2000*
			Current Standard		
NE	0.23	0.25	0.31	0.25	0.26
MA	2.06	1.70	1.66	1.62	1.69
SA	3.09	3.62	3.53	3.73	3.80
ENC	3.50	3.96	3.79	3.69	3.61
ESC	2.64	2.44	2.38	2.28	2.00
WNC	1.32	1.61	1.58	1.70	1.83
WSC	0.11	0.95	1.52	1.73	1.91
NM	0.12	0.09	0.13	0.18	0.24
SM	0.34	0.23	0.30	0.28	0.28
PA	0.20	0.31	0.31	0.33	0.29
National	13.6	15.2	15.5	15.8	15.9
			80% Standard		
NE		0.24	0.29	0.22	0.22
MA		1.68	1.59	1.52	1.53
SA		3.93	3.82	3.82	3.68
ENC		4.36	3.96	3.64	3.22
ESC		2.44	2.39	2.28	2.05
WNC		1.43	1.37	1.53	1.76
WSC		0.78	1.02	1.12	1.14
NM		0.07	0.07	0.09	0.11
SM		0.20	0.20	0.20	0.18
PA		0.27	0.25	0.21	0.16
National		15.4	15.0	14.3	14.1
			90% Standard		
NE		0.23	0.25	0.18	0.18
MA		1.65	1.49	1.36	1.29
SA		3.88	3.60	3.49	3.24
ENC		4.30	3.90	3.58	3.17
ESC		2.44	2.35	2.25	2.00
WNC		1.43	1.35	1.49	1.69
WSC		0.74	0.84	0.79	0.76
NM		0.06	0.05	0.06	0.07
SM		0.19	0.17	0.16	0.13
PA		0.26	0.22	0.19	0.13
National		15.2	14.3	13.6	12.7

*Million metric tons per year.
Source: Teknekron, Inc.

APPENDIX 2

Results of EPA Analyses Completed in April, 1978 and August, 1978, Summarizing the Costs and Effects of Alternative NSPS for Coal-Fired Power Plants

Source: Environmental Protection Agency, "Electric Utility Steam Generating Units: Proposed Standards of Performance and Announcement of Public Hearing on Proposed Standards, *Federal Register*, Vol. 43, No. 182, September 19, 1978, p. 42165.

Table 1: Comparison of Assumptions. April 1978 and August 1978

Assumption	*April*	*August*
Growth rates	1975–1985: 5.8%/yr 1985–1995: 5.5%	1975–1985: 4.8%/yr 1985–1995: 4.0%
Nuclear capacity	1985: 108 GW 1990: 177 1995: 302	1985: 97 GW 1990: 167 1995: 230
Oil prices ($ 1975)	1985: $13/bbl 1990: $13 1995: $13	1985: $15/bbl 1990: $20 1995: $28
General inflation rate	5.5%/yr	5.5%/yr
Annual emissions @ 0.5 floor	0.5 lb SO_2/million Btu	0.32 lb SO_2/million Btu
Coal transportation	Increases at general inflation rate	Increases at general inflation rate plus 1%
Coal mining labor costs	Increases at general inflation rate	Increases at general inflation rate plus 1%
Miscellaneous	A number of miscellaneous changes were made between the April 1978 study and the August 1978 study. These changes were either corrections or refinements of values used in the April study. Examples of these changes included revisions to the level of SIP control assumed in the model, revisions to the scrubbing costs, changes in the assumptions regarding industrial coal consumption, and changes to the coal supply curves used in the April study.	

Table 2: Summary of National 1990 SO_2 Emissions from Utility Boilers[a] (million tons)

		Level of Control									
	1975 Actual	Current Standards		Full Control		210 ng/J		Partial Control 290 ng/J		340 ng/J	
Plant Category		APR	AUG	APR	AUG	APR	AUG	APR	AUG	APR	AUG
SIP/NSPS Plants[b]	—	16.8	16.0	17.2	16.2	16.9	16.2	—	16.1	16.7	16.1
New Plants[c]	—	4.2	4.4	1.5	1.2	2.1	1.3	—	1.5	3.3	1.8
Oil/Gas Plants	—	2.3	1.1	2.5	1.4	2.3	1.2	—	1.2	2.3	1.2
Total National Emissions	18.6	23.3	21.4	21.1	18.9	21.3	18.8	—	18.9	22.3	19.1
Total Coal Capacity (GW)	205	465	451	444	428	460	439	—	440	460	444

Source: Background Information for Proposed SO_2 Emission Standards—Supplement, EPA 450/2-78-007a-1, Chapters 2 and 3, August 1978.

[a]Results of EPA analyses completed in April 1978 and August 1978.

[b]Plants subject to existing state regulations or the current NSPS of 1.2 lb SO_2/million Btu.

[c]Plants subject to the revised standards.

Table 3. Summary of 1990 Regional SO_2 Emissions for Utility Boilers[a] (million tons)

	Level of Control										
	1975 Actual	*Current Standards*		*Full Control*				*Partial Control*			
						210 ng/J		*290 ng/J*		*340 ng/J*	
		APR	*AUG*	*APR*	*AUG*	*APR*	*AUG*	*APR*	*AUG*	*APR*	*AUG*
Total National Emissions	18.6	23.3	21.4	21.1	18.9	21.3	18.8	—	18.9	22.3	19.1
Regional Emissions											
East[b]	9.1	10.8	10.2	9.7	9.0	9.6	9.0	—	8.9	10.2	9.0
Midwest[c]	8.8	8.7	7.8	8.5	7.6	8.4	7.6	—	7.6	8.6	7.6
West South Central[d]	0.2	2.6	2.3	1.8	1.5	2.0	1.4	—	1.5	2.3	1.6
West[e]	0.5	1.3	1.3	1.1	0.8	1.2	0.9	—	0.9	1.3	1.0
Total Coal Capacity (GW)	205	465	451	444	428	460	439	—	440	460	444

Source: *Background Information for Proposed SO_2 Emission Standards—Supplement*, EPA 450/2-78-0071-1, Chapters 2 and 3, August 1978.

[a]Results of EPA analyses completed in April 1978 and August 1978.

[b]New England, Middle Atlantic, South Atlantic, and East South Central Census Regions.

[c]East North Central and West North Central Census Regions.

[d]West South Central Census Region.

[e]Mountain and Pacific Census Regions.

Table 4: Summary of 1990 SO_2 Emissions by Plants Subject to the Proposed Standards: August 1978 Analysis

	Level of Control				
	Current Standards	*Full Control*	*Partial Control 210 ng/J*	*290 ng/J*	*340 ng/J*
East[a]					
Total New Plant Emissions (million tons)	2.1	0.7	0.7	0.7	0.8
Coal Consumption (10^{15} Btu)	3.47	3.41	3.43	3.48	3.47
Emission Factor (#S/10^6 Btu)[b]	0.60	0.21	0.21	0.22	0.23
Midwest[c]					
Total New Plant Emissions (million tons)	0.60	0.2	0.2	0.2	0.2
Coal Consumption (10^{15} Btu)	1.17	0.79	0.80	0.81	0.81
Emission Factor (#S/10^6 Btu)[b]	0.48	0.21	0.21	0.23	0.26
West South Central[d]					
Total New Plant Emissions (million tons)	1.2	0.2	0.3	0.4	0.5
Coal Consumption (10^{15} Btu)	1.93	1.67	1.97	1.96	1.95
Emission Factor (#S/10^6/Btu)[b]	0.60	0.14	0.14	0.18	0.24
West[e]					
Total New Plant Emissions (million tons)	0.6	0.1	0.2	0.2	0.3
Coal Consumption (10^{15} Btu)	1.25	1.19	1.18	1.19	1.24
Emission Factor (#S/10^6/Btu)[b]	0.40	0.09	0.14	0.19	0.24

Source: Background Information for Proposed SO_2 Emission Standards—Supplement, EPA 450/2-78-007a-1, Chapter 3, August 1978.

[a]New England, Middle Atlantic, South Atlantic, and East South Central Census Regions.

[b]Ratios may not be obtained exactly from figures shown here due to rounding.

[c]East North Central and West North Central Census Regions.

[d]West South Central Census Region.

[e]Mountain and Pacific Census Regions.

Table 5: Summary of 1995 SO_2 Emissions by Plants Subject to the Proposed Standards: August 1978 Analysis

	Level of Control				
	Current Standards	Full Control	Partial Control 210 ng/J	290 ng/J	340 ng/J
East[a]					
Total New Plant Emissions (million tons)	4.0	1.3	1.3	1.4	1.5
Coal Consumption (10^{15} Btu)	6.73	6.39	6.47	6.49	6.62
Emission Factor (#S/10^6 Btu)[b]	0.60	0.21	0.21	0.21	0.22
Midwest[c]					
Total New Plant Emissions (million tons)	1.2	0.4	0.4	0.5	0.5
Coal Consumption (10^{15} Btu)	2.21	1.94	1.92	1.99	2.00
Emission Factor (#S/10^6 Btu)[b]	0.53	0.21	0.21	0.23	0.26
West South Central[d]					
Total New Plant Emissions (million tons)	1.6	0.4	0.4	0.5	0.7
Coal Consumption (10^{15} Btu)	2.63	2.77	2.73	2.70	2.68
Emission Factor (#S/10^6 Btu)[b]	0.60	0.15	0.15	0.19	0.25
West[e]					
Total New Plant Emissions (million tons)	1.1	0.2	0.3	0.4	0.5
Coal Consumption (10^{15} Btu)	2.28	2.32	2.29	2.27	2.27
Emission Factor (#S/10^6 Btu)[b]	0.44	0.09	0.13	0.19	0.22

Source: *Background Information for Proposed SO_2 Emission Standards—Supplement*, EPA 450/2-78-007a-1, Chapter 3, August 1978.

[a]New England, Middle Atlantic, South Atlantic, and East South Central Census Region.

[b]Ratios may not be obtained exactly from figures shown here due to rounding.

[c]East North Central and West North Central Census Regions.

[d]West South Central Census Region.

[e]Mountain and Pacific Census Regions.

TABLE 6: Summary of impacts on fuels in 1990[a]

	Level of Control										
	1975 Actual	Current Standards		Full Control		Partial Control 210 ng/J		290 ng/J		340 ng/J	
		APR	AUG	APR	AUG	APR	AUG	APR	AUG	APR	AUG
U.S. Coal Production (million tons)											
East	396	441	465	467	449	464	450	—	450	418	449
Midwest	151	298	275	375	318	353	316	—	294	307	290
West	100	1027	785	870	736	938	752	—	779	1055	784
TOTAL	647	1767	1525	1711	1502	1755	1517	—	1523	1780	1523
Western coal shipped east (million tons)	21	455	149	299	118	346	117	—	147	429	152
Oil/gas consumption in power plants (million bbl/day)	3.1	3.0	1.2	3.3	1.5	3.1	1.4	—	1.4	3.1	1.4

Source: *Background Information for Proposed SO_2 Emission Standards - Supplement*, EPA 450/2-78-007a-1, Chapter 2 & 3, August 1978.

[a]Results of EPA analyses completed in April 1978 and August 1978.

TABLE 7: Summary of 1990 Economic Impacts[a]

	Level of Control									
	Current Standards		Full Control		210 ng/J		Partial Control 290 ng/J		340 ng/J	
	APR	AUG	APR	AUG	APR	AUG	APR	AUG	APR	AUG
Average monthly residential bills ($/month)	45.31	43.89	46.39	44.22	46.20	44.48	—	44.38	45.47	44.38
Incremental Utility capital expenditures, cumulative 1976–1990 ($ billions)	—	—	10	0	15	8	—	4	3	5
Incremental Annualized cost ($ billions)	—	—	2.0	1.9	1.3	1.7	—	1.3	0.3	1.1
Incremental Cost of SO_2 Reduction ($/ton)	—	—	885	754	640	642	—	511	303	485

Source: *Background Invormation for Proposed SO_2 Emission Standards - Supplement*, EPA 450/2-78-007a-1, Chapters 2 & 3, August 1978.

[a]Results of EPA analyses completed in April 1978 and August 1978.

TABLE 8: Summary of 1995 Impacts: August 1978 Analysis

	Level of Control					
	1975 Actual	*Current Standards*	*Full Control*		*Partial Control*	
				210 ng/J	*290 ng/J*	*340 ng/J*
National Emissions (million tons)	18.6	23.3	18.5	18.5	18.7	19.0
New Plant Emissions[a] (million tons)	—	7.9	2.4	2.5	2.8	3.2
U.S. Coal Production (million tons)	647	1865	1865	1858	1868	1866
Western Coal Shipped East (million tons)	21	210	130	133	190	196
Oil/Gas Consumption (million bbl/day)	3.1	0.8	0.9	0.9	0.9	0.9
Incremental Cumulative Capital Expenditures (1975 $ billion)	—	—	32	26	20	19
Incremental Annualized Cost (1975 $ billion)	—	—	2.6	2.3	2.0	1.9
Average Monthly Residential Bill (1975 $/month)	—	46.34	46.22	46.13	46.12	46.10
Total Coal Capacity (GW)	198	587	580	580	580	580

Source: *Background Information for Proposed SO_2 Emission Standards-Supplement*, EPA 450/2-78-007a-1, Chapter 3, August 1978.

[a]Plants subject to the revised standards.

APPENDIX 3

COMMENT BY ENVIRONMENTAL ORGANIZATIONS ON THE SEPTEMBER 1978 *FEDERAL REGISTER* PROPOSAL CONCERNING NEW SOURCE PERFORMANCE STANDARDS FOR COAL-FIRED POWER PLANTS

I. *Richard E. Ayres, David Doniger, Veronica M. Kun for Natural Resources Defense Council and Robert J. Rauch, David Lennett for Environmental Defense Fund,* Comments on Proposed Standards of Performance for New Electric Generating Units, *January 15, 1979. (In the formal comments submitted to EPA on the September proposal.)*

When presented in page after page of tables, the output of this exercise appears formidably precise. Like many a computer exercise, however, the quality of the output is only as good as the quality of the input, and the care with which the model is constructed and interpreted. In this case, there are substantial weaknesses in all these factors which render many of the conclusions suspect or misleading, usually in a way that tends to suggest that stronger proposals are less desirable than they would otherwise appear. . . .The model's conclusions are largely determined by two questionable assumptions: the assumed price of oil and the assumption of "substitutability."

The most obviously surprising conclusion of the modeling exercise is that the predicted costs and total emissions reduction from all the various proposals analyzed are remarkably similar, regardless of whether full scrubbing or partial scrubbing is modeled. Even more curiously, EPA's supplemental report suggests that partial scrubbing would be even more expensive than a standard requiring full scrubbing. In fact, the largest difference between the two projections, in percentage terms, lies in the projected use of oil, where the model suggests that a full scrubbing policy would result in the utilities using considerably more oil than would a policy of partial scrubbing.

These predictions are so at odds with common sense that they cry out for explanation. Upon examination, the reason for these anomalies emerges rapidly. They exist because of the assumptions the modelers have made about the price of oil in the future, and because they have assumed that the utilities will choose between building new coal-fired power plants, on the one hand, and "stretching out" the lives of old, poorly controlled oil- and coal-fired power plants or building new oil turbines, on the other, entirely on the basis of the assumed cost difference between these alternatives at the time they make their decision. Implicit in this assumption is also the assumption that the utilities will be entirely free to substitute these alternatives for building coal-fired power plants conforming to the NSPS. Given these assumptions, the model predicts that the small difference in cost between uncontrolled and controlled coal-fired power plants will be sufficient to induce the utilities to rely far more on old plants and new oil-fired turbines.

So long as the model assumes oil prices within the range chosen by the modelers and adopts the hypothesis that old oil- and coal-fired capacity can be freely substituted for new coal-fired capacity, it will minimize the projected advantage in reducing emissions of full scrubbing over partial scrubbing, and minimize the apparent cost-effectiveness of full scrubbing, while maximizing the apparent difference in oil consumption associated with the two. . . . If the assumed oil price is too low, or if substitutions will be prevented by economic risk-averse behavior by utilities or by noneconomic factors, or could be prevented by conscious government intervention, then the modelers have grossly understated the difference in environmental benefits between full and partial scrubbing alternatives. We think both assumptions are highly questionable.

Despite the apparent complexity of the model, its conclusions are based on an extreme oversimplification of reality. The model excluded complexities, such as secondary effects of the environment, energy, and cost that would result from the political impact of adopting a sliding scale, particularly in the West where a "number of states now have emission control requirements for new coal-fired power plants that are considerably more demanding than the requirements proposed in DOE's and the utilities' sliding scale proposals."

Similarly, the promulgation of a lax federal "sliding scale" could reasonably be expected to provoke a new round of bitter confrontation between the utilities and the citizens who fought for strengthened federal requirements to preserve their air quality.

II. *Richard E. Ayres and David P. Doniger, "New Source Standards for Power Plants II: Consider the Law,"* Harvard Environmental Law Review, *Vol. 3:48, 1979.*

In the case of three of these variables—oil prices, scrubber costs, and the effects of coal variability—the Agency's modelers have chosen assumptions that exaggerate the apparent costs and minimize the apparent effectiveness of the more environmentally sound proposals.

The modelers have assumed essentially static oil prices through 1995. We know of no respected authority on energy who would support this assumption. The modelers have exaggerated the effects of coal variability on the costs of scrubbers (as we discuss below in Section VC.3), because they have assumed the utilities will cope with this phenomenon by building expensive excess scrubber capacity (and, in the case of UARG's modelers, excess electric generating capacity), rather than by adopting less expensive means to eliminate the problem. Finally, they have assumed that the real costs of scrubbers will remain the same through 1995. This assumption is highly unrealistic.

The test we draw from the (court cases about NSPS) is that if EPA declines to require any technical feasible emission reduction, the agency must show that the costs of such reductions would [be] "grossly disproportionate" to their benefits . . . The cases uniformly reject the claim that EPA must set standards on the basis of formal cost-benefit analysis. In view of the grave known and potential harm from power plant emissions, and the fact that the increase in the average household electricity bill probably will be less than thirty cents per month, even for the highest degree of scrubbing, this showing would be very difficult to make.

Ultimately, however, we do not believe "cost-effectiveness" is a sound basis for deciding what level of control the NSPS should require. Cost-effectiveness analysis is a technique for choosing the most efficient way to accomplish a given goal, but it does not help determine what the goal should be. In order to know what level of emission reduction should be required, one would need more accurate measures of the health and environmental damage the emissions would cause, accurate measures of the costs of controls, and adequate means of comparing these incommensurable benefits and costs. Aware that we lack these capabilities, Congress and the courts have wisely refrained from requiring the NSPS decision to be based on formal "cost-benefit" analysis. Cost-effectiveness analysis is even less appropriate, since it generates figures that, even when accurate, have no context whatsoever.

APPENDIX 4

Partial Results of Joint Environmental Protection Agency/ Department of Energy Analyses Completed in November, 1978, of Costs and Effects of Alternative NSPS for Coal-Fired Power Plants

Source: Environmental Protection Agency, "Standards of Performance for New Stationary Sources: Electric Utility Steam Generating Units; Additional Information," *Federal Register*, Vol. 43, No. 237, December 8, 1978, pp. 57837–57856.

TABLE A4-1: Comparison of Assumptions, August 1978 and November 1978

Assumption	*August*	*November*	
Growth rates	1975–1985 4.8%/yr	same	
	1985–1995 4.0%	same	
Nuclear capacity	1985 97 GW	same	
	1990 165	1990 165 GW	
	1995 230	1995 228	
Oil prices ($ 1975)		*Run 1*	*Run 2*
	1985 $15/bbl	1985 $12.90/bbl	$12.30/bbl
	1990 $20	1990 $16.40	$13.20
	1995 $28	1995 $21.00	$14.90
General inflation rate	5.5%/yr	same	
Coal transportation	Increases at general inflation rate plus 1%	same	
Coal-mining labor costs	U.M.W. settlement and 1% real increase thereafter	same	
Capital charge rate	10%	12.5%	
Cost reporting basis	1975 dollars	1978 dollars	
Coal cleaning credit	No	Yes	
FGD costs	—	Refined	

TABLE A4-2: National 1990 SO_2 Emissions from Utility Boilers with 100 Percent FGD Reliability[a] (million tons)—Full Control

Plant Category	*1975 Actual*	*Current Standards*	*90% Control with 340 ng/J Limit*	*90% Control with 240 ng/J Limit*	*90% Control across Scrubber*
SIP/NSPS plants[b]	—	16.1	16.4	16.5	16.5
New plants[c]	—	3.6	1.6	1.2	1.1
Oil plants	—	1.8	1.9	1.9	1.9
Total national emissions	18.6	21.5	19.9	19.6	19.5
Total coal capacity (GW)	198	406	391	390	390

[a]Results of joint EPA/DOE analyses completed in May 1979 based on oil prices of $12.90, $16.40, and $21.00/bbl in years 1985, 1990, and 1995, respectively.

[b]Plants subject to existing state regulations or the current NSPS of 1.2 lb. SO_2/mill. Btu.

[c]Plants subject to the revised standards.

TABLE A4-3: National 1990 SO_2 Emissions from Utility Boilers with 100 Percent FGD Reliability[a] (million tons)—Partial Control

Plant Category	*1975 Actual*	*Current Standards*	*240 ng/J Emission Limit*	*90% Control with 330 ng/J Max. Control*	*Sliding Scale with 90% Control in West*	*240 ng/J Standard with 90% Control in West*
SIP/NSPS plants[b]	—	16.1	16.5	16.4	16.5	16.4
New plants[c]	—	3.6	1.6	2.2	1.3	1.4
Oil plants	—	1.8	1.9	1.8	1.9	1.9
Total national emissions	18.6	21.5	20.0	20.4	19.7	19.7
Total coal capacity (GW)	198	406	399	401	394	398

[a]Results of joint EPA/DOE analyses completed in May 1979 based on oil prices of $12.90, $16.40, and $21.00/bbl in the years 1985, 1990, and 1995, respectively.

[b]Plants subject to existing state regulations or the current NSPS of 1.2 lb. SO_2/mill. Btu.

[c]Plants subject to the revised standards.

TABLE A4-4: National 1990 SO_2 Emissions from Utility Boilers with 100 Percent FGD Reliability[a] (million tons)

			Level of Control			
Plant Category	*1975 Actual*	*Current Standards*	*90% Control with 150 ng/J Max. Control*	*90% Control with 240 ng/J Max. Control*	*Sliding Scale with 390 ng/J Limit*	*95% Control with 520 ng/J Limit*
SIP/NSPS plants[b]	—	16.1	16.6	16.5	16.2	16.4
New plants[c]	—	3.6	1.3	1.9	3.1	0.8
Oil plants	—	1.8	1.9	1.9	1.8	1.8
Total national emissions	18.6	21.5	19.8	20.3	21.1	19.0
Total coal capacity (GW)	198	406	396	399	401	400

[a]Results of EPA analyses completed in May 1979 based on oil prices of $12.90, $16.40, and $21.00/bbl in the years 1985, 1990, and 1995, respectively.

[b]Plants subject to existing state regulations or the current NSPS of 1.2 lb. SO_2/mill. Btu.

[c]Plants subject to the revised standards.

TABLE A4-5: National 1990 SO_2 Emissions from Utility Boilers with 90 Percent FGD Reliability[a] (million tons)

			Full Control		Partial Control
Plant Category	*1975 Actual*	*Current Standards*	*90% Control with 240 ng/J Limit*	*90% Control Across Scrubber*	*240 ng/J Emission Limit*
SIP/NSPS plants[b]	—	16.1	16.5	16.5	16.6
New plants[c]	—	3.6	1.2	1.1	1.6
Oil plants	—	1.8	2.0	2.0	1.9
Total national emissions	18.6	21.5	19.7	19.6	20.1
Total coal capacity (GW)	198	406	388	388	397

[a]Results of joint EPA/DOE analyses completed in November 1978 based on oil prices of $12.90, $16.40, and $21.00/bbl in the years 1985, 1990, and 1995, respectively.

[b]Plants subject to existing state regulations or the current NSPS of 1.2 lb. SO_2/mill. Btu.

[c]Plants subject to the revised standards.

TABLE A4-6: Regional 1990 SO_2 Emissions from Utility Boilers with 100 Percent FGD Reliability[a] (million tons)—Full Control

	1975 Actual	*Current Standards*	*90% Control with 340 ng/J Limit*	*90% Control with 240 ng/J Limit*	*90% Control across Scrubber*
Total national emissions	18.6	21.5	19.9	19.6	19.5
Regional emissions					
East[b]	—	7.0	6.6	6.4	6.4
East South Central[c]	—	3.2	3.0	2.9	2.9
Midwest[d]	—	5.4	5.5	5.5	5.5
Great Plains[e]	—	2.4	2.3	2.3	2.3
West South Central[f]	—	2.2	1.6	1.6	1.6
West[g]	—	1.3	0.9	0.9	0.9
Total coal capacity (GW)	198	406	391	390	390

[a]Results of joint EPA/DOE analyses completed in November 1978 based on oil prices of $12.90, $16.40, and $21.00/bbl in the years 1985, 1990, and 1995, respectively.

[b]New England, Middle Atlantic, and South Atlantic Census Regions.

[c]East South Central Census Region.

[d]East North Central Census Region.

[e]West North Central Census Region.

[f]West South Central Census Region.

[g]Mountain and Pacific Census Regions.

TABLE A4-7: Regional 1990 SO_2 Emissions from Utility Boilers with 100 Percent FGD Reliability[a] (million tons)—Partial Control

	1975 Actual	Current Standards	240 ng/J Emission Limit	90% Control with 330 ng/J Max. Control	Sliding Scale with 90% Control in West	240 ng/J Standard with 90% Control in West
Total national emissions	18.6	21.5	20.0	20.4	19.7	19.7
Regional emissions						
East[b]	—	7.0	6.4	6.6	6.4	6.4
East South Central[c]	—	3.2	2.9	3.0	2.9	2.9
Midwest[d]	—	5.4	5.5	5.5	5.4	5.4
Great Plains[e]	—	2.4	2.4	2.4	2.5	2.4
West South Central[f]	—	2.2	1.8	1.9	1.7	1.8
West[g]	—	1.3	1.0	1.1	0.9	0.9
Total coal capacity (GW)	198	406	399	401	394	398

[a]Results of joint EPA/DOE analyses completed in November 1978 based on oil prices of $12.90, $16.40, and $21.00/bbl in the years 1985, 1990, and 1995, respectively.

[b]New England, Middle Atlantic, and South Atlantic Census Regions.

[c]East South Central Census Region.

[d]East North Central Census Region.

[e]West North Central Census Region.

[f]West South Central Census Region.

[g]Mountain and Pacific Census Regions.

TABLE A4-8: Regional 1990 SO_2 Emissions from Utility Boilers with 90 Percent FGD Reliability[a] (million tons)

			Full Control		Partial Control
	1975 Actual	*Current Standards*	*90% Control with 240 ng/J Limit*	*90% Control Across Scrubber*	*240 ng/J Emission Limit*
Total national emissions	18.6	21.5	19.7	19.6	20.1
Regional emissions					
East[b]	—	7.0	6.4	6.4	6.5
East South Central[c]	—	3.2	2.9	2.9	2.9
Midwest[d]	—	5.4	5.6	5.5	5.5
Great Plains[e]	—	2.4	2.3	2.3	2.4
West South Central[f]	—	2.2	1.6	1.6	1.8
West[g]	—	1.3	0.9	0.9	1.1
Total coal capacity (GW)	198	406	388	388	397

[a]Results of joint EPA/DOE analyses completed in November 1978 based on oil prices of $12.90, $16.40, and $21.00/bbl in the years 1985, 1990, and 1995, respectively.

[b]New England, Middle Atlantic and South Atlantic Census Regions.

[c]East South Central Census Region.

[d]East North Central Census Region.

[e]West North Central Census Region.

[f]West South Central Census Region.

[g]Mountain and Pacific Census Regions.

TABLE A4-9: Regional 1990 SO_2 Emissions from Utility Boilers with 100 Percent FGD Reliability[a] (million tons)

			Partial Control			
	1975 Actual	*Current Standards*	*90% Control with 150 ng/J Max. Control*	*90% Control with 240 ng/J Max. Control*	*Sliding Scale with 390 ng/J Limit*	*95% Control with 520 ng/J Limit*
Total national emissions	18.6	21.5	19.8	20.3	21.1	19.0
Regional emissions						
East[b]	—	7.0	6.4	6.5	6.9	6.2
East South Central[c]	—	3.2	2.9	3.0	3.2	2.8
Midwest[d]	—	5.4	5.6	5.6	5.5	5.5
Great Plains[e]	—	2.4	2.3	2.4	2.4	2.2
West South Central[f]	—	2.2	1.6	1.7	2.0	1.5
West[g]	—	1.3	1.0	1.0	1.2	0.8
Total coal capacity (GW)	198	406	396	399	401	400

[a]Results of joint EPA/DOE analyses completed in November 1978 based on oil prices of $12.90, $16.40, and $21.00/bbl in the years 1985, 1990, and 1995, respectively.

[b]New England, Middle Atlantic, and South Atlantic Census Regions.

[c]East South Central Census Region.

[d]East North Central Census Region.

[e]West North Central Census Region.

[f]West South Central Census Region.

[g]Mountain and Pacific Census Regions.

TABLE A4-10: Impacts on Fuels in 1990 with 100 Percent FGD Reliability[a]—Full Control

	1975 Actual	*Current Standards*	*90% Control with 340 ng/J Limit*	*90% Control with 240 ng/J Limit*	*90% Control across Scrubber*
U.S. coal production (million tons)					
Appalachia	—	419	399	398	398
Midwest	—	348	390	390	390
Northern Great Plains	—	463	458	460	461
West	—	211	188	188	188
Total	647	1441	1435	1436	1437
Western Coal Shipped East (million tons)	21	84	62	64	64
Oil Consumption by Power Plants (million bbl/day)					
Power Plants	—	1.8	2.0	2.1	2.1
Coal Transportation	—	0.1	0.2	0.1	0.1
Total	3.1	1.9	2.2	2.2	2.2

[a]Results of joint EPA/DOE analyses completed in November 1978 based on oil prices of $12.90, $16.40, and $21.00/bbl in the years 1985, 1990, and 1995, respectively.

TABLE A4-11: Impacts on Fuels in 1990 with 100 Percent FGD Reliability[a]—Partial Control

	1975 Actual	*Current Standards*	*240 ng/J Emission Limit*	*90% Control with 330 ng/J Max. Control*	*Sliding Scale with 90% Control in West*	*240 ng/J Standard with 90% Control in West*
U.S. coal production (millon tons)						
Appalachia	—	419	412	415	411	411
Midwest	—	348	377	373	367	377
Northern Great Plains	—	463	455	451	465	459
West	—	211	200	202	196	197
Total	647	1441	1444	1441	1439	1444
Western coal shipped East (million tons)	21	84	62	67	85	64
Oil consumption by power plants (million bbl/day)						
Power plants	—	1.8	1.9	1.9	2.0	1.9
Coal transportation	—	0.1	0.1	0.1	0.1	0.1
Total	3.1	1.9	2.0	2.0	2.1	2.0

[a]Results of joint EPA/DOE analyses completed in November 1978 based on oil prices of $12.90, $16.40, and $21.00/bbl in the years 1985, 1990, and 1995, respectively.

TABLE A4-12: Impacts on Fuels in 1990 with 90 Percent FGD Reliability[a]

			Full Control		*Partial Control*
	1975 Actual	*Current Standards*	*90% Control with 40 ng/J Limit*	*90% Control Across Scrubber*	*240 ng/J Emission Limit*
U.S. coal production (million tons)					
Appalachia	—	419	398	398	411
Midwest	—	348	387	387	374
Northern Great Plains	—	463	460	459	450
West	—	211	183	184	202
Total	647	1441	1428	1428	1437
Western coal shipped east (million tons)	21	84	65	64	63
Oil consumption by power plants (million bbl/day)					
Power plants	—	1.8	2.1	2.1	1.9
Coal transportation	—	0.1	0.1	0.1	0.1
Total	3.1	1.9	2.2	2.2	2.1

[a]Results of joint EPA/DOE analyses completed in November 1978 based on oil prices of $12.90, $16.40, and $21.00/bbl in the years 1985, 1990, and 1995, respectively.

TABLE A4-13: Impacts on Fuels in 1990 with 100 Percent FGD Reliability[a]

			Partial Control			
	1975 Actual	Current Standards	90% Control with 150 ng/J Max. Control	90% Control with 240 ng/J Max. Control	Sliding Scale with 390 ng/J Limit	95% Control with 520 ng/J Limit
U.S. coal production (million tons)						
Appalachia	—	419	403	406	410	403
Midwest	—	348	392	385	372	394
Northern Great Plains	—	463	445	451	465	465
West	—	211	199	199	198	188
Total	647	1441	1439	1441	1445	1450
Western Coal Shipped East (million tons)	21	84	60	60	76	62
Oil Consumption by power plants (million bbl/day)						
Power plants	—	1.8	1.9	1.9	1.9	1.9
Coal transportation	—	0.1	0.1	0.1	0.1	0.1
Total	3.1	1.9	2.0	2.0	2.0	2.0

[a]Results of joint EPA/DOE analyses completed in November 1978 based on oil prices of $12.90, $16.40, and $21.00/bbl in the years 1985, 1990, and 1995, respectively.

TABLE A4-14: Economic Impacts in 1990 with 100 Percent FGD Reliability[a] (1978 $)—Full Control

	Current Standards	*90% Control with 340 ng/J Limit*	*90% Control with 240 ng/J Limit*	*90% Control Across Scrubber*
Average monthly residential bills ($/month)	50.52	51.32	51.35	51.37
Incremental utility capital expenditures, cumulative 1976–1990 ($ billions)	—	2.5	2.4	2.5
Incremental annualized cost ($ billions)	—	1.9	2.0	2.1
Incremental cost of SO_2 reduction ($/ton)	—	1236	1041	1010

[a]Results of joint EPA/DOE analyses completed in November 1978 based on oil prices of $12.90, $16.40, and $21.00/bbl in the years 1985, 1990, and 1995, respectively.

TABLE A4-15: Economic Impacts in 1990 with 100 Percent FGD Reliability[a] (1978 $)—Partial Control

	Current Standards	*240 ng/J Emission Limit*	*90% Control with 330 ng/J Max. Control*	*Sliding Scale with 90% Control in West*	*240 ng/J Standard with 90% Control in West*
Average monthly residential bills ($/month)	50.52	51.18	51.04	51.30	51.28
Incremental utility capital expenditures, cumulative 1976–1990 ($ billions)	—	5.5	4.8	4.1	6.0
Incremental annualized cost ($ billions)	—	1.5	1.1	1.8	1.7
Incremental cost of SO_2 reduction ($/ton)	—	955	1099	1014	958

[a]Results of EPA analyses completed in November 1978 based on oil prices of $12.90, $16.40, and $21.00/bbl in the years 1985, 1990, and 1995, respectively.

TABLE A4-16: Economic Impacts in 1990 with 100 Percent FGD Reliability[a] (1978 $)

		Partial Control			
	Current Standards	90% Control with 150 ng/J Max. Control	90% Control with 240 ng/J Max. Control	Sliding Scale with 390 ng/J Limit	95% Control[b] with 520 ng/J Limit
Average monthly residential bills ($/month)	50.52	51.35	51.16	50.90	51.12
Incremental utility capital expenditures, cumulative 1976–1990 ($ billions)	—	6.4	5.5	5.1	5.5
Incremental annualized cost ($ billions)	—	1.8	1.4	0.7	1.3
Incremental cost of SO_2 reduction ($/ton)	—	1080	1090	2045	524

[a]Results of EPA analyses completed in November 1978 based on oil prices of $12.90, $16.40, and $21.00/bbl in the years 1985, 1990, and 1995, respectively.

[b]Modified FGD cost functions used.

TABLE A4-17: Economic Impacts in 1990 with 90 Percent FGD Reliability[a] (1978 $)

		Full Control		*Partial Control*
	Current Standards	*90% Control with 240 ng/J Limit*	*90% Control Across Scrubber*	*240 ng/J Emission Limit*
Average monthly residential bills ($/month)	50.52	51.34	51.36	51.23
Incremental utility capital expenditures, cumulative 1976–1990 ($ billions)	—	—	0.20	4.70
Incremental annualized cost ($ billions)	—	2.10	2.10	1.60
Incremental cost of SO_2 reduction ($/ton)	—	1171	1103	1115

[a]Results of EPA analyses completed in November 1978 based on oil prices of $12.90, $16.40, and $21.00/bbl in the years 1985, 1990, and 1995, respectively.

TABLE A4-18: Summary of 1995 Impacts with 100 Percent FGD Reliability[a]—Full Control

	1975 Actual	*Current NSPS*	*90% Control with 340 ng/J Limit*	*90% Control with 240 ng/J Limit*	*90% Control Across Scrubber*
National emissions (million tons SO_2)	18.6	23.7	20.5	19.8	19.5
New plant emissions[b] (million tons SO_2)	—	7.1	3.2	2.3	2.0
U.S. coal production (million tons)	647	1779	1765	1767	1768
Western coal shipped east (million tons)	21	122	59	77	77
Oil consumption (million bbl/day)	3.1	1.3	1.7	1.7	1.7
Incremental cumulative capital expenditures (1978 $ billion)	—	—	4.4	4.4	5.0
Incremental annualized cost (1978 $ billion)	—	—	4.1	4.3	4.3
Average monthly residential Bill (1978 $/month)	—	53.03	54.31	54.37	54.40
Total Coal Capacity (GW)	198	552	521	520	520

[a]Results of EPA analyses completed in November 1978 based on oil prices of $12.90, $16.40, and $21.00/bbl in the years 1985, 1990, and 1995, respectively.

[b]Plants subject to the revised standards.

TABLE A4-19: Summary of 1995 Impacts with 100 Percent FGD Reliability[a]—Partial Control

	1975 Actual	*Current NSPS*	*240 ng/J Emission Limit*	*90% Control with 330 ng/J Max. Control*	*Sliding Scale with 90% Control in West*	*240 ng/J Standard with 90% Control in West*
National emissions (million tons SO_2)	18.6	23.70	20.30	21.20	19.90	20.0
New plant emissions[b] (million tons SO_2)	—	7.10	3.20	4.30	2.50	2.80
U.S. coal production (million tons)	647	1779	1765	1762	1769	1767
Western coal shipped east (million tons)	21	122	73	79	92	73
Oil consumption (million bbl/day)	3.1	1.30	1.60	1.50	1.60	1.60
Incremental cumulative capital expenditures (1978 $ billion)	—	—	5.90	3.70	9.00	7.40
Incremental annualized cost (1978 $ billion)	—	—	3.30	2.60	4.00	3.70
Average monthly residential bill (1978 $/month)	—	53.03	54.08	53.84	54.34	54.24
Total coal capacity (GW)	198	552	534	537	531	533

[a]Results of EPA analyses completed in November 1978 based on oil prices of $12.90, $16.40, and $21.00/bbl in the years 1985, 1990, and 1995, respectively.

[b]Plants subject to the revised standards.

TABLE A4-20: Summary of 1995 Impacts with 90 Percent FGD Reliability[a]

			Full Control		*Partial Control*
	1975 Actual	*Current NSPS*	*90% Control with 40 ng/J Limit*	*90% Control Across Scrubber*	*240 ng/J Emission Limit*
National emissions (million tons SO_2)	18.60	23.70	20.10	19.80	20.50
New plant emissions[b] (million tons SO_2)	—	7.10	2.40	2.10	3.20
U.S. coal production (million tons)	647	1779	1761	1762	1760
Western coal shipped east (million tons)	21	122	79	81	74
Oil consumption (million bbl/day)	3.1	1.30	1.80	1.80	1.60
Incremental cumulative capital expenditures (1978 $ billion)	—	—	3.50	4.20	5.20
Incremental annualized cost (1978 $ billion)	—	—	4.60	4.60	3.50
Average monthly residential bill (1978 $/month)	—	53.03	54.46	54.47	54.14
Total coal capacity (GW)	198	552	520	520	534

[a]Results of EPA analyses completed in November 1978 based on oil prices of $12.90, $16.40, and $21.00/bbl in the years 1985, 1990, and 1995, respectively.

[b]Plants subject to the revised standards.

TABLE A4-21: Summary of 1995 Impacts with 90 Percent FGD Reliability[a]

			Partial Control			
	1975 Actual	*Current NSPS*	*90% Control with 150 ng/J Max. Control*	*90% Control with 240 ng/J Max. Control*	*Sliding Scale with 390 ng/J Limit*	*95% Control with 520 ng/J Limit[c]*
National emissions (million tons SO_2)	18.60	23.70	20.00	20.80	22.80	18.20
New plant emissions[b] (million tons SO_2)	—	7.10	2.60	3.70	6.00	1.60
U.S. coal production (million tons)	647	1779	1762	1765	1765	1799
Western coal shipped East (million tons)	21	122	68	69	83	54
Oil consumption (million bbl/day)	3.10	1.30	1.6	1.60	1.50	1.50
Incremental cumulative Captial Expenditures (1978 $ billion)	—	—	8.1	5.70	3.60	7.20
Incremental annualized Cost (1978 $ billion)	—	—	4.0	3.10	1.90	0.80
Average monthly residential bill (1978 $/month)	—	53.03	54.33	54.04	53.64	53.31
Total coal capacity (GW)	198	552	531	534	537	542

[a]Results of EPA analyses completed in November 1978 based on oil prices of $12.90, $16.40, and $21.00/bbl in the years 1985, 1990, and 1995, respectively.

[b]Plants subject to the revised standards.

[c]Modified FGD cost functions used.

APPENDIX 5

ENVIRONMENTAL PROTECTION AGENCY'S DESCRIPTION OF REGULATORY ANALYSIS AND RESULTS OF JOINT EPA/DOE ANALYSES COMPILED IN MAY, 1979

Regulatory Analysis

Executive Order 12044 (March 24, 1978), whose objective is to improve Government regulations, requires executive branch agencies to prepare regulatory analysis for regulations that may have major economic consequences. EPA has extensively analyzed the costs and other impacts of these regulations. These analyses, which meet the criteria for preparation of a regulatory analysis, are contained within the preamble to the proposed regulations (43 FR 42154), the background documentation made available to the public at the time of proposal (see STUDIES, 43 FR 42171), this preamble, and the additional background information document accompanying this action ("Electric Utility Steam Generating Units, Background Information for Promulgated EmissionStandards," EPA-450/3–79–021). Due to the volume of this material and its continual development over a period of 2–3 years, it is not practical to consolidate all analyses into a single document. The following discussion gives a summary of the most significant alternatives considered. The rationale for the action taken for each pollutant being regulated is given in a previous section.

Source: Environmental Protection Agency, "New Stationary Sources Performance Standards; Electric Utility Steam Generating Units," *Federal Register*, Vol. 44, No. 113, June 11, 1979, pp. 33601–33609.

In order to determine the appropriate form and level of control for the standards, EPA has performed extensive analysis of the potential national impacts associated with the alternative standards. EPA employed economic models to forecast the structure and operating characteristics of the utility industry in future years. These models project the environmental, economic, and energy impacts of alternative standards for the electric utility industry. The major analytical efforts took place in three phases as described below.

Phase 1. The initial effort comprised a preliminary analysis completed in April 1978 and a revised assessment completed in August 1978. These analyses were presented in the September 19, 1978 **Federal Register** proposal (43 FR 42154). Corrections to the September proposal package and additional information was published on November 27, 1978 (43 FR 55258). Further details of the analyses can be found in "Background Information for Proposed SO_2 Emission Standards—Supplement," EPA 450/2–78–007a–1.

Phase 2. Following the September 19 proposal, the EPA staff conducted additional analysis of the economic, environmental, and energy impacts associated with various alternative sulfur dioxide standards. As part of this effort, the EPA staff met with representatives of the Department of Energy, Council of Economic Advisors, Council on Wage and Price Stability, and others for the purpose of reexamining the assumptions used for the August analysis and to develop alternative forms of the standard for analysis. As a result, certain assumptions were changed and a number of new regulatory alternatives were defined. The EPA staff again employed the economic model that was used in August to project the national and regional impacts associated with each alternative considered.

The results of the phase 2 analysis were presented and discussed at the public hearings in December and were published in the **Federal Register** on December 8, 1978 (43 FR 7834).

Phase 3. Following the public hearings, the EPA staff continued to analyze the impacts of alternative sulfur dioxide standards. There were two primary reasons for the continuing analysis. First, the detailed analysis (separate from the economic modeling) of regional coal production impacts pointed to a need to investigate a range of higher emission limits.

Secondly, several comments were received from the public regarding the potential of dry sulfur dioxide scrubbing systems. The phase 1 and phase 2 analyses had assumed that utilities would use wet scrubbers only. Since dry scrubbing costs substantially less then wet scrubbing, adoption of the dry technology would substantially change the economic, energy, and environmental impacts of alternative sulfur dioxide standards. Hence, the phase 3 analysis focused on the impacts of alternative standards under a range of emission ceilings assuming both wet technology and the adoption of dry scrubbing for applications in which it is technically and economically feasible.

Impacts Analyzed

The environmental impacts of the alternative standards were examined by projecting pollutant emissions. The emissions were estimated nationally and by geographic region for each plant type, fuel type, and age category. The EPA staff also evaluated the waste products that would be generated under alternative standards.

The economic and financial effects of the alternatives were examined. This assessment included an estimation of the utility capital expenditures for new plant and pollution control equipment as well as the fuel costs and operating and maintenance expenses associated with the plant and equipment. These costs were examined in terms of annualized costs and annual revenue requirements. The impact on consumers was determined by analyzing the effect of the alternatives on average consumer costs and residential electric bills. The alternatives were also examined in terms of cost per ton of SO_2 removal. Finally, the present value costs of the alternatives were calculated.

The effects of the alternative proposals on energy production and consumption were also analyzed. National coal use was projected and broken down in terms of production and consumption by geographic region. The amount of western coal shipped to the Midwest and East was also estimated. In addition, utility consumption of oil and natural gas was analyzed.

Major Assumptions

Two types of assumptions have an important effect on the results of the analysis. The first group involves the model structure and characteristics. The second group includes the assumptions used to specify future economic conditions.

The utility model selected for this analysis can be characterized as a cost minimizing economic model. In meeting demand, it determines the most economic mix of plant capacity and electric generation for the utility system, based on a consideration of construction and operating costs for new plants and variable costs for existing plants. It also determines the optimum operating level for new and existing plants. This economic-based decision criteria should be kept in mind when analyzing the model results. These criteria imply, for example, that all utilities base decisions on lowest costs and that neutral risk is associated with alternative choices.

Such assumptions may not represent the utility decision making process in all cases. For example, the model assumes that a utility bases supply decisions on the cost of constructing and operating new capacity versus the cost of operating existing capacity. Environmentally, this implies a tradeoff between emissions from new and old sources. The cost minimization assumption implies that in meeting the standard a new power plant will fully scrub high-sulfur coal if this option is cheaper than fully or partially scrubbing low-sulfur coal. Often the model will have to make such a decision, especially in the Midwest where utilities can choose between burning local high-sulfur or imported western low-sulfur coal. The assumption of risk neutrality implies that a utility will always choose the low-cost option. Utilities, however, may perceive full scrubbing as involving more risks and pay a premium to be able to partially scrub the coal. On the other hand, they may perceive risks associated with long-range transportation of coal, and thus opt for full control even though partial control is less costly.

The assumptions used in the analyses to represent economic conditions in a given year have a significant impact on the final results reached. The major assumptions used in the analyses are shown in Table 1 and the significance of these parameters is summarized below.

The growth rate in demand for electric power is very important since this rate determines the amount of new capacity which will be needed and thus directly affects the emission estimates and the projections of pollution control costs. A high electric demand growth rate results in a larger emission reduction associated with the proposed standards and also results in higher costs.

The nuclear capacity assumed to be installed in a given year is also important to the analysis. Because nuclear power is less expensive, the model will predict construction of new nuclear plants rather than new coal plants. Hence, the nuclear capacity assumption affects the amount of new coal capacity which will be required to meet a given electric demand level. In practice, there are a number of constraints which limit the amount of nuclear capacity which can be constructed, but for this study, nuclear capacity was specified approximately equal to the moderate growth projections of the Department of Energy.

The oil price assumption has a major impact on the amount of predicted new coal capacity, emissions, and oil consumption. Since the model makes generation decisions based on cost, a low oil price relative to the cost of building and operating a new coal plant will result in more oil-fired generation and less coal utilization. This results in less new coal capacity which reduces capital costs but increases oil consumption and fuel costs because oil is more expensive per Btu than coal. This shift in capacity utilization also affects emissions, since an existing oil plant generally has a higher emission rate than a new coal plant even when only partial control is allowed on the new plant.

Coal transportation and mine labor rates both affect the delivered price of coal. The assumed transportation rate is generally more important to the predicted consumption of low-sulfur coal (relative to high-sulfur coal), since that is the coal type which is most often shipped long distances. The assumed mining labor cost is more important to eastern coal costs and production estimates since this coal production is generally much more labor intensive than western coal.

Because of the uncertainty involved in predicting future economic conditions, the Administrator anticipated a large number of comments from the public regarding the modeling assumptions. While the Administrator would have liked to analyze each scenario under a range of assumptions for each critical parameter, the number of modeling inputs made such an approach impractical. To decide on the best assumptions, and to limit the number of sensitivity runs, a joint working group was formed. The group was comprised of representatives from the Department of Energy, Council of Economic Advisors, Council on Wage and Price Stability, and others. The group reviewed model results to date, identified the key inputs, specified the assumptions, and identified the critical parameters for which the degree of uncertainty was such that sensitivity analyses should be performed. Three months of study resulted in a number of changes which are reflected in Table 1 and discussed below. These assumptions were used in both the phase 2 and phase 3 analyses.

After more evaluation, the joint working group concluded that the oil prices assumed in the phase 1 analysis were too high. On the other hand, no firm guidance was available as to what oil prices should be used. In view of this, the working group decided that the best course of action was to use two sets of oil prices which reflect the best estimates of those governmental entities concerned with projecting oil prices. The oil price sensitivity analysis was part of the phase 2 analysis which was distributed at the public hearing. Further details are available in the draft report, "Still Further Analysis of Alternative New Source Performance Standards for New Coal-Fired Power Plants (docket number IV-A-5)." The analysis showed that while the variation in oil price affected the magnitude of emissions, costs, and energy impacts, price variation had little effect on the relative impacts of the various NSPS alternatives tested. Based on this conclusion, the higher oil price was selected for modeling purposes since it paralleled more closely the middle range projections by the Department of Energy.

TABLE 1: Key Modeling Assumptions

Assumption	
Growth rates	1975–1985: 4.8%/yr.
	1985–1995: 4.0%.
Nuclear capacity	1985: 97 GW.
	1990: 165.
	1995: 228.
Oil prices ($ 1975)	1985: $12.90/bbl.
	1990: $16.40.
	1995: $21.00.
Coal transportation	1% per year real increase.
Coal mining labor costs	U.M.W. settlement and 1% real increase thereafter.
Capital charge rate	12.5% for pollution control expenditures.
Cost reporting basis	1978 dollars.
FGD costs	No change from phase 2 analysis except for the addition of dry scrubbing systems for certain applications.
Coal cleaning credit	5%–35% SO_2 reduction assumed for high sulfur bituminous coals only.
Bottom ash and fly ash content	No credit assumed.

Reassessment of the assumptions made in the phase 1 analysis also revealed that the impact of the coal washing credit had not been considered in the modeling analysis. Other credits allowed by the September proposal, such as sulfur removed by the pulverizers or in bottom ash and fly ash, were determined not to be significant when viewed at the national and regional levels. The coal washing credit, on the other hand, was found to have a significant effect on predicted emissions levels and, therefore, was factored into the analysis.

As a result of this reassessment, refinements also were made in the fuel gas desulfurization (FGD) costs assumed. These refinements include changes in sludge disposal costs, energy penalties calculated for reheat, and module sizing. In addition, an error was corrected in the calculation of partial scrubbing costs. These changes have resulted in relatively higher partial scrubbing costs when compared to full scrubbing.

Changes were made in the FGD availability assumption also. The phase 1 analysis assumed 100 percent availability of FGD systems. This assumption, however, was in conflict with EPA's estimates on module availability. In view of this, several alternatives in the phase 2 analysis were modeled at lower system availabilities. The assumed availability was consistent with a 90 percent availability for individual modules when the system is equipped with one spare. The analysis also took into consideration the emergency by-pass provisions of the proposed regulation. The analysis showed that lower reliabilities would result in somewhat higher emissions and costs for both the partial and full control cases. Total coal capacity was slightly lower under full control and slightly higher under partial control. While it was postulated that the lower reliability assumption would produce greater adverse impacts on full control than on partial control options, the relative differences in impacts were found to be insignificant. Hence, the working group discarded the reliability issue as a major consideration in the analyzing of national impacts of full and partial control options. The Administrator still believes that the newer approach better reflects the performance of well designed, operated, and maintained FGD systems.However, in order to expedite the analysis, all subsequent alternatives were analyzed with an assumed system reliability of 100 percent.

Another adjustment to the analysis was the incorporation of dry SO_2 scrubbing systems. Dry scrubbers were assumed to be available for both new and retrofit applications. The costs of these systems were estimated by EPA's Office of Research and Development based on pilot plant studies and contract prices for systems currently under construction. Based on economic analysis, the use of dry scrubbers was assumed for low-sulfur coal (less than 1290 ng/J or 3 lb. SO_2/million Btu) applications in which the control requirement was 70 percent or less. For higher sulfur content coals, wet scrubbers were assumed to be more economical. Hence, the scenarios characterized as using "dry" costs contain a mix of wet and dry technology whereas the "wet" scenarios assume wet scrubbing technology only.

Additional refinements included a change in the capital charge rate for pollution control equipment to conform to the Federal tax laws on depreciation, and the addition of 100 billion tons of coal reserves not previously accounted for in the model.

Finally, a number of less significant adjustments were made. These included adjustments in nuclear capacity to reflect a cancellation of a plant, consideration of oil consumption in transporting coal, and the adjustment of costs to 1978 dollars rather than 1975 dollars. It should be understood that all reported costs include the costs of complying with the proposed particulate matter standard and NO_X standards, as well as the sulfur dioxide alternatives. The model does not incorporate the Agency's PSD regulations nor forthcoming requirements to protect visibility.

Public Comments

Following the September proposal, a number of comments were received on the impact analysis. A great number focused on the model inputs, which were reviewed in detail by the joint working group. Members of the joint working group represented a spectrum of expertise (energy, jobs, environment, inflation, commerce). The following paragraphs discuss only those comments addressed to parts of the analysis which were not discussed in the preceding section.

One commenter suggested that the costs of complying with State Implementation Plan (SIP) regulations and prevention of significant deterioration requirements should not be charged to the standards. These costs are not charged to the standards in the analyses. Control requirements under PSD are based on site specific, case-by-case decisions for which the standards serves as a minimum level of control. Since these judgments cannot be forecasted accurately, no additional control was assumed by the model beyond the requirements of these standards. In addition, the cost of meeting the various SIP regulations was included as a base cost in all the scenarios modeled. Thus, any forecasted cost differences among alternative standards reflect differences in utility expenditures attributable to changes in the standards only.

Another commenter believed that the time horizon for the analysis (1990/1995) was too short since most plants on line at that time will not be subject to the revised standard. Beyond 1995, our data show that many of the power plants on line today will be approaching retirement age. As utilization of older capacity declines, demand will be picked up by newer, better controlled plants. As this replacement occurs, national SO_2 emissions will begin to decline. Based on this projection, the Administrator believes that the 1990-1995 time frame will represent the peak years for SO_2 emissions and is, therefore, the relevant time frame for this analysis.

Use of a higher general inflation rate was suggested by one commenter. A distinction must be made between general inflation rates and real cost escalation. Recognizing the uncertainty of future inflation rates, the EPA staff conducted the economic analysis in a manner that minimized reliance on this assumption. All construction, operating, and fuel costs were expressed as constant year dollars and therefore the analyis is not affected by the inflation rate. Only real cost escalation was included in the economic analysis. The inflation rates will have an impact on the present value discount rate chosen since this factor equals the inflation rate plus the real discount rate. However, this impact is constant across all scenarios and will have little impact on the conclusions of the analysis.

Another commenter opposed the presentation of economic impacts in terms of monthly residential electric bills, since this treatment neglects the impact of higher energy costs to industry. The Administrator agrees with this comment and has included indirect consumer impacts in the analysis. Based on results of previous analysis of the electric utility industry, about half of the total costs due to pollution control are felt as direct increases in residential electric bills. The increased costs also flow into the commercial and industrial sectors where they appear as increased costs of consumer goods. Since the Administrator is unaware of any evidence of a multiplier effect on these costs, straight cost pass through was assumed. Based on this analysis, the indirect consumer impacts (Table 5) were concluded to be equal to the monthly residential bills ("Economic and Financial Impacts of Federal Air and Water Pollution Controls on the Electric Utility Industry," EPA–230/3–76/013, May 1976).

One utility company commented that the model did not adequately simulate utility operation since it did not carry out hour-by-hour dispatch of generating units. The model dispatches by means of load duration curves which were developed for each of 35 demand regions across the United States. Development of these curves took into consideration representative daily load curves, traditional utility reserve margins, seasonal demand variations, and historical generation data. The Administrator believes that this approach is adequate for forecasting long-term impacts since it plans for meeting short-term peak demand requirements.

Summary of Results

The final results of the analysis are presented in Tables 2 through 5 and discussed below. For the three alternative standards presented, emission limits and percent reduction requirements are 30-day rolling averages, and each standard was analyzed with a particulate standard of 13 ng/J (0.03 lb/million Btu) and the proposed NO_X standards. The full control option was specified as a 520 ng/J (1.2 lb/million Btu) emission limit with a 90 percent reduction in potential SO_2 emissions. The other options are the same as full control except when the emissions to the atmosphere are reduced below 260 ng/J (0.6 lb/million Btu) in which case the minimum percent reduction requirement is reduced. The variable control option requires a 70 percent minimum reduction and the partial control option has a 33 percent minimum reduction requirement. The impacts of each option were forecast first assuming the use of wet scrubbers only and then assuming introduction of dry scrubbing technology. In contrast to the September proposal which focused on 1990 impacts, the analytical results presented today are for the year 1995. The Administrator believes that 1995 better represents the differences among alternatives since more new plants subject to the standard will be on line by 1995. Results of the 1990 analyses are available in the public record.

Wet Scrubbing Results

The projected SO_2 emissions from utility boilers are shown by plant type and geographic region in Tables 2 and 3. Table 2 details the 1995 national SO_2 emissions resulting from different plant types and age groups. These standards will reduce 1995 SO_2 emissions by about 3 million tons per year (13 percent) as compared to the current standards. The emissions from new plants directly affected by the standards are reduced by up to 55 percent. The emission reduction from new plants is due in part to lower emission rates and in part to reduced coal consumption predicted by the model. The reduced coal consumption in new plants results from the increased cost of constructing and operating new coal plants due to pollution controls. With these increased costs, the model predicts delays in construction of new plants and changes in the utilization of these plants after start-up. Reduced coal consumption by new plants is accompanied by higher utilization of existing plants and combustion turbines. This shift causes increased emissions from

TABLE 2: National 1995 SO_2 Emissions From Utility Boilers[a] [Million tons]

Plant category	1975 actual	Level of control[b]: Current standards		Full control		Partial control 33% minimum		Variable control 70% minimum	
		Wet[d]	*Dry*[e]	*Wet*	*Dry*	*Wet*	*Dry*	*Wet*	*Dry*
SIP/NSPS Plants[c]		15.5	15.8	16.0	16.2	15.9	16.2	16.0	16.1
New Plants[f]		7.1	7.0	3.1	3.1	3.6	3.4	3.3	3.1
Oil Plants		1.0	1.0	1.4	1.4	1.3	1.2	1.3	1.2
Total National Emissions	18.6	23.7	23.8	20.6	20.7	20.8	20.9	20.6	20.5
Total Coal Capacity (GW)	205	552	554	521	520	534	537	533	537
Sludge generated (million tons dry)		23	27	55	56	43	39	50	41

[a]Results of EPA analyses completed in November 1978 based on oil prices of $12.90, $16.40, and $21.00/bbl in the years 1985, 1990, and 1995, respectively.

[b]With 520 ng/J maximum emission limit.

[c]Plants subject to existing State regulations or the current NSPS of 1.2 lb SO_2/million BTU.

[d]Based on wet SO_2 scrubbing costs.

[e]Based on dry SO_2 scrubbing costs where applicable.

[f]Plants subject to the revised standards.

TABLE 3: Regional 1995 SO_2 Emissions From Utility Boilers[a] [Million tons]

	Level of control[b]								
	1975 actual	*Current standards*		*Full control*		*Partial control 33% minimum*		*Variable control 70% minimum*	
		Wet[c]	*Dry*[d]	*Wet*	*Dry*	*Wet*	*Dry*	*Wet*	*Dry*
Total National Emissions	18.6	23.7	23.8	20.6	20.7	20.8	20.9	20.6	20.5
Regional Emissions:									
East[e]		11.2	11.2	10.1	10.1	9.8	9.8	9.8	9.7
Midwest[f]		8.1	8.3	7.9	7.9	7.9	8.0	7.9	8.0
West South Central[g]		2.6	2.6	1.7	1.7	1.8	1.8	1.8	1.7
West[h]		1.7	1.7	0.9	0.9	1.2	1.2	1.1	1.1
Total Coal Capacity (GW)	205	552	554	521	520	534	537	533	537

[a]Results of EPA analyses completed in November 1978 based on oil prices of $12.90, $16.40, and $21.00/bbl in the years 1985, 1990, and 1995, respectively.

[b]With 520 ng/J maximum emission limit.

[c]Based on wet SO_2 scrubbing costs.

[d]Based on dry SO_2 scrubbing costs where applicable.

[e]New England, Middle Atlantic, South Atlantic, and East South Central Census Regions.

[f]East North Central and West North Central Census Regions.

[g]West South Central Census Region.

[h]Mountain and Pacific Census Regions.

TABLE 4: Impacts on Fuels in 1995[a]

	Level of control[b]								
	1975 actual	Current standards		Full control		Partial control 33% minimum		Variable control 70% minimum	
		Wet[c]	Dry[d]	Wet	Dry	Wet	Dry	Wet	Dry
U.S. Coal Production (million tons):									
Appalachia	396	489	524	463	465	475	486	470	484
Midwest	151	404	391	487	488	456	452	465	450
Northern Great Plains	54	655	630	633	628	622	576	632	602
West	46	230	222	182	180	212	228	203	217
Total	647	1,778	1,767	1,765	1,761	1,765	1,742	1,770	1,752
Western Coal Shipped East (millon tons)	21	122	99	59	55	68	59	71	70
Oil Consumption by Power Plant (million bbl/day):									
Power Plants		1.2	1.2	1.6	1.6	1.4	1.4	1.4	1.4
Coal Transportation		0.2	0.2	0.2	0.2	0.2	0.2	0.2	0.2
Total	3.1	1.4	1.4	1.8	1.8	1.6	1.6	1.6	1.6

[a]Results of EPA analyses completed in November 1978 based on oil prices of $12.90, $16.40, and $21.00/bbl in the years 1985, 1990, and 1995, respectively.

[b]With 520 ng/J maximum emission limit.

[c]Based on wet SO_2 scrubbing costs.

[d]Based on dry SO_2 scrubbing where applicable.

existing coal- and oil-fired plants, which partially offsets the emission reductions achieved by new plants subject to the standard.

Projections of 1995 regional SO_2 emissions are summarized in Table 3. Emissions in the East are reduced by about 10 to 13 percent as compared to predictions under the current standards, whereas Midwestern emissions are reduced only slightly. The smaller reductions in the Midwest are due to a slow growth of new coal-fired capacity. In general, introductions of coal-fired capacity tends to reduce emissions since new coal plants replace old coal- and oil-fired units which have higher emission rates. The greatest emission reduction occurs in the West and West South Central regions where significant growth is expected and today's emissions are relatively low. For these two regions combined, the full control option reduces emissions by 40 percent from emission levels under the current standards, while the partial and variable options produce reductions of about 30 percent.

Table 4 illustrates the effect of the proposed standards on 1995 coal production, western coal shipped east, and utility oil and gas consumption. National coal production is predicted to triple by 1995 under all the alternative standards. This increased demand raises production in all regions of the country as compared to 1975 levels. Considering these major increases in national production, the small production variations among the alternatives are not large. Compared to production under the current standards, production is down somewhat in the West, Northern Great Plains, and Appalachia, while production is up in the Midwest. These shifts occur because of the reduced economic advantage of low-sulfur coals under the revised standards. While three times higher than 1975 levels, western coal shipped east is lower under all options than under the current standards.

Oil consumption in 1975 was 1.4 million barrels per day. The 3.1 million barrels per day figure for 1975 consumption in Table 4 includes utility natural gas consumption (equivalent of 1.7 million barrels per day) which the analysis assumed would be phased out by 1990. Hence, in 1995, the 1.4 million barrel per day projection under current standards reflects retirement of existing oil capacity and offsetting increases in consumption due to gas-to-oil conversions.

Oil consumption by utilities is predicted to increase under all the options. Compared to the current standards, increased consumption is 200,000 barrels per day under the partial and variable options and 400,000 barrels per day under full control. Oil consumption differences are due to the higher costs of new coal plants under these standards, which causes a shift to more generation from existing oil plants and combustion turbines. This shift in generation mix has important implications for the decision-making process, since the only assumed constraint to utility oil use was the price. For example, if national energy policy imposes other constraints which phase out or stabilize oil use for electric power generation, then the differences in both oil consumption and oil plant emissions (Table 2) across the various standards will be mitigated. Constraining oil consumption, however, will spread cost differences among standards.

The economic effects in 1995 are shown in Table 5. Utility capital expenditures increase under all options as compared to the $770 billion estimated to be required through 1995 in the absence of a change in the standard. The capital estimates in Table 5 are increments over the expenditures under the current standard and include both plant capital (for new capacity) and pollution control expenditures. As shown in Table 2, the model estimates total industry coal capacity to be about 17 GW (3 percent) greater under the non-uniform control options. The cost of this extra capacity makes the total utility capital expenditures higher under the partial and variable options, than under the full control option, even though pollution control capital is lower.

Annualized cost includes levelized capital charges, fuel costs, and operation and maintenance costs associated with utility equipment. All of the options cause an increase in annualized cost over the current standards. This increase ranges from a low of $3.2 billion for partial control to $4.1 billion for full control, compared to the total utility annualized costs of about $175 billion.

The average monthly bill is determined by estimating utility revenue requirements which are a function of capital expenditures, fuel costs, and operation and maintenance costs. The average bill is predicted to increase only slightly under any of the options, up to a maximum 3-percent increase shown for full control. Over half of the large total increase in the average monthly bill over 1975 levels ($25.50 per month) is due to a significant increase in the amount of electricity used by each customer. Pollution control expenditures, including those to meet the current standards, account for about 15 percent of the increase in the cost per kilowatt-hour while the remainder of the cost increase is due to capital intensive capacity expansion and real escalations in construction and fuel cost.

Indirect consumer impacts range from $1.10 to $1.60 per month depending on the alternative selected. Indirect consumer impacts reflect increases in consumer prices due to the increased energy costs in the commercial and industrial sectors.

The incremental costs per ton of SO_2 removal are also shown in Table 5. The figures are determined by dividing the change in annualized cost by the change in annual emissions, as compared to the current standards. These ratios are a measure of the cost effectiveness of the

TABLE 5: 1995 Economic Impacts[a] [1978 dollars]

	Level of control[b]							
	Current standards		Full control		Partial control 33% minimum		Variable control 70% minimum	
	Wet[c]	*Dry*[d]	*Wet*	*Dry*	*Wet*	*Dry*	*Wet*	*Dry*
Average Monthly Residential Bills ($/month)	$53.00	$52.85	$54.50	$54.45	$54.15	$53.95	$54.30	54.05
Indirect Consumer Impacts ($/month)			1.50	1.60	1.15	1.10	1.30	1.20
Incremental Utility Capital Expenditures, Cumulative 1976–1995 ($ billions)			4	5	6	−3	10	−1
Incremental Annualized Cost ($ billions)			4.1	4.4	3.2	3.0	3.6	3.3
Present Value of Incremental Utility Revenue Requirements ($ billions)			41	45	32	31	37	33
Incremental Cost of SO_2 Reduction ($/ton)			1,322	1,428	1,094	1,012	1,163	1,036

[a]Results of EPA analyses completed in May 1979 based on oil prices of $12.90, $16.40, and $21.00/bbl in the years 1985, 1990, and 1995, respectively.

[b]With 520 ng/J maximum emission limit.

[c]Based on wet SO_2 scrubbing costs.

[d]Based on dry SO_2 scrubbing costs where applicable.

options, where lower ratios represent a more efficient resource allocation. All the options result in higher cost per ton than the current standards with the full control option being the most expensive.

Another measure of cost effectiveness is the average dollar-per-ton cost at the plant level. This figure compares total pollution control cost with total SO_2 emission reduction for a model plant. This average removal cost varies depending on the level of control and the coal sulfur content. The range for full control is from $325 per ton on high-sulfur coal to $1,700 per ton on low-sulfur coal. On low-sulfur coals, the partial control cost is $2,000 per ton, and the variable cost is $1,700 per ton.

The economic analyses also estimated the net present value cost of each option. Present value facilitates comparison of the options by reducing the streams of capital, fuel, and operation and maintenance expenses to one number. A present value estimate allows expenditures occurring at different times to be evaluated on a similar basis by discounting the expenditures back to a fixed year. The costs chosen for the present value analysis were the incremental utility revenue requirements relative to the current NSPS. These revenue requirements most closely represent the costs faced by consumers. Table 5 shows that the present value increment for 1995 capacity is $41 billion for full control, $37 billion for variable control, and $32 billion for partial control.

Dry Scrubbing Results

Tables 2 through 5 also show the impacts of the options under the assumption that dry SO_2 scrubbing systems penetrate the pollution control market. These analyses assume that utilities will install dry scrubbing systems for all applications where they are technologically feasible and less costly than wet systems. (See earlier discussion of assumptions.)

The projected SO_2 emissions from utility boilers are shown by plan type and geographic region in Tables 2 and 3. National emission projections are similar to the wet scrubbing results. Under the dry control assumption, however, the variable control option is predicted to have the lowest national emissions primarily due to lower oil plant emissions relative to the full control option. Partial control produces more emissions than variable control because of higher emissions from new plants. Compared to the current standards, regional emission impacts are also similar to the wet scrubbing projections. Full control results in the lowest emissions in the West, while variable control results in the lowest emissions in the East. Emissions in the Midwest and West South Central are relatively unaffected by the options.

Inspection of Tables 2 and 3 shows that with the dry control assumption the current standard, full control, and partial control cases produce slightly higher emissions than the corresponding wet control cases. This is due to several factors, the most important of which is a shift in the generation mix. This shift occurs because dry scrubbers have lower capital costs and higher variable costs than wet scrubbers and, therefore, the two systems have different effects on the plant utilization rates. The higher variable costs are due primarily to transportation charges on intermediate to low sulfur coal which must be used with dry scrubbers. The increased variable cost of dry controls alters the dispatch order of existing plants so that older, uncontrolled plants operate at relatively higher capacity factors than would occur under the wet scrubbing assumption, hence increasing total emissions. Another factor affecting emissions is utility coal selection which may be altered by differences in pollution control costs.

Table 4 shows the effect to the proposed standards on fuels in 1995. National coal production remains essentially the same whether dry or wet controls are assumed. However, the use of dry controls causes a slight reallocation in regional coal production, except under a full control option where dry controls cannot be applied to new plants. Under the variable and partial options Appalachian production increases somewhat due to greater demand for intermediate sulfur coals while Midwestern coal production declines slightly. The non-uniform options also result in a small shifting in the western regions with Northern Great Plains production declining and production in the rest of West increasing. The amount of western coal shipped east under the current standard is reduced from 122 million to 99 million tons (20% decrease) due to the increased use of eastern intermediate sulfur coals for dry scrubbing applications. Western coal shipped east is reduced further by the revised standards, to a low of 55 million tons under full control. Oil impacts under the dry control assumption are identical to the wet control cases, with full control resulting in increased consumption of 200 thousand barrels per day relative to the partial and variable options.

The 1995 economic effects of these standards are presented in Table 5. In general, the dry control assumption results in lower costs. However, when comparing the dry control costs to the wet control figures it must be kept in mind that the cost base for comparison, the current standards, is different under the dry control and wet control assumptions. Thus, while the incremental costs of full control are higher under the dry scrubber assumption the total costs of meeting the standard are lower than if wet controls were used.

The economic impact figures show that when dry controls are assumed the cost savings associated with the variable and partial options is significantly increased over the wet control cases. Relative to full control the partial control option nets a savings of $1.4 billion in annualized costs which equals a $14 billion net present value savings. Variable control results in a $1.1 billion annualized cost savings which is a savings of $12 billion in net present value. These changes in utility costs affect the average residential bill only slightly, with partial control resulting in a savings of $.50 per month and variable control savings of $.40 per month on the average bill, relative to full control.

Conclusions

One finding that has been clearly demonstrated by the two years of analysis is that lower emission standards on new plants do not necessarily result in lower national SO_2 emissions when total emissions from the entire utility system are considered. There are two reasons for this finding. First, the lowest emissions tend to result from strategies that encourage the construction of new coal capacity. This capacity, almost regardless of the alternative analyzed, will be less polluting than the existing coal- or oil-fired capacity that it replaces. Second, the higher cost of operating the new capacity (due to higher pollution costs) may cause the newer, cleaner plants to be utilized less than they would be under a less stringent alternative. These situations are demonstrated by the analyses presented here.

The variable control option produces emissions that are equal to or lower than the other options under both the wet and dry scrubbing assumptions. Compared to full control, variable control is predicted to result in 12 GW to 17 GW more coal capacity. This additional capacity replaces dirtier existing plants and compensates for the slight increase in emissions from new plants subject to the standards, hence causing emissions to be less than or equal to full control emissions depending on scrubbing cost assumption (i.e., wet or dry). Partial control and variable control produce about the same coal capacity, but the additional 300 thousand ton emission reduction from new plants causes lower total emissions under the variable option. Regionally, all the options produce about the same emissions in the Midwest and West South Central regions. Full control produces 200 thousands tons less emissions in the West than the varible option and 300 thousand tons less than partial control. But the variable and partial options produce between 200 and 300 thousand tons less emissions in the East.

The variable and partial control options have a clear advantage over full control with respect to costs under both the wet and dry scrubbing assumptions. Under the dry assumption, which the Administrator believes represents the best prediction of utility behavior, variable control saves about $1.1 billion per year relative to full control and partial control saves an additional $0.3 billion.

All the options have similar impacts on coal production especially when considering the large increase predicted over 1975 production levels. With respect to oil consumption, however, the full control option causes a 200,000 barrel per day increase as compared to both the partial and variable options.

Based on these analyses, the Administrator has concluded that a non-uniform control strategy is best considering the environmental, energy, and economic impacts at both national and regional levels. Compared to other options analyzed, the variable control standard presented above achieves the lowest emissions in an efficient manner and will not disrupt local or regional coal markets. Moreover, this option avoids the 200 thousand barrel per day oil penalty which has been predicted under a number of control options. For these reasons, the Administrator believes that the variable control option provides the best balance of national environmental, energy, and economic objectives.

ABOUT THE AUTHOR

ELIZABETH H. HASKELL is a consultant on environmental policy and public administration. She has held senior positions in Congress and the federal Executive Branch, and has served eight years as a member of Virginia's air pollution regulatory board. Her many books and articles include *State Environmental Management: Case Studies of Nine States* (Praeger, 1973).